The Learning Handbook Series

Understanding Basic Electricity

A Non-Technical Introduction For Everyone

by

Irwin Math

ARTISAN IDEAS

Published by Artisan North America, 2026.

Written by Irwin Math
Layout and cover design: Andrea James Blaho
Technical Editor: Leonardo Barsantini

ISBN 9781737604440
Printed in the UK
Artisan Ideas is an imprint of **Artisan North America, Inc.**
Info@ArtisanIdeas.com

Artisan North America
753 Valley Road
Watchung, NJ 07069-6120
USA

Trade enquiries: enquiries@mare-nostrum.co.uk.

To VISHA,
My Primary Source of Power

Contents

List of Illustrations 5

Foreword 6

1) **A Brief Introduction**
What Is Electricity?...Circuits 7

2) **Some Basic Terms**
Volts...Amperes...Ohms...Ohm's Law 12

3) **Sources of Electricity**
Batteries...Generators 18

4) **Tools of the Trade**
Multimeter...Wire Strippers...Pliers 25

5) **Basic Electrical Components**
Switches...Connectors...Basic Power Distribution 30

6) **Common Electrical Symbols and Diagrams**
Electrical Schematic Symbols 36

7) **Reading and Understanding Schematic Diagrams**
Reading Schematics 39

8) **Electrical Protection Devices**
Fuses...Circuit Breakers...Consumer Units 45

9) **Basic Electrical Troubleshooting**
Continuity Measurements...DC and AC Measurements 50

10) **Summing It All Up**
What Have We Learned?...Final Comments 55

11) **Some Basic Electrical Repairs** 58

Appendix 64

Author Biography 70

List of Illustrations

1. Simple Water-Based Shower System 8
2. Simple Electrical-Based Lighting System 8
3. Cutaway View of an Electrical Wire 9
4. Opposites Attract 12
5. Ohm's Law 14
6. Average Power Required by Some Common Devices 16
7. Internal View of a 9-Volt Battery 18
8. Common Battery 19
9. Simple Electrical Circuit 20
10. Typical Digital Multimeter 21
11. Explanation of Generator Operation 22
12. Close-up of Multimeter Dial 25
13. Common Wire Strippers 28
14. Typical Electrical Tools 28
15. Several Electrical Switches 30
16. Common Switch 30
17. Barrel, Battery and Crimp-Type Connectors 31
18. AC Power Line Outlets and Plugs 32
19. The Protective Earth Connection 32
20. Common Power Distribution Wire 33
21. Front and Back Views of a 3-Pin to 2-Pin Adapter 34
22. Schematic Diagram of Figure 3 36
23. Basic Electrical Schematic Symbols 37
24. Common Table Lamp Schematic Diagram 39
25. Two-Switch Light Control Schematic 40
26. Simple Vacuum Cleaner Schematic Diagram 41
27. Basic Self Defrosting Refrigerator Schematic Diagram 41
28. Cross Section of an "Old Fashioned Fuse" 45
29. Cross Section of Modern Fuses 46
30. Typical Consumer Unit 46
31. Typical Self-Resetting Consumer Unit 47
32. Typical RCD-Protected Socket Outlet 48
33. Multimeter Continuity Tester Settings 50
34. Multimeter DC Voltage Settings 52
35. Multimeter AC Voltage Settings 53
36. Common Electrical Warning Signs 56
37. Typical Electrical Terminals and Splices 60
38. Fuse and Consumer Unit Status 61
39. Common AC Mains Plug Wiring 66

Foreword

Understanding Basic Electricity was written to give an ordinary reader of any age a general idea of what electricity is, how it is produced, and how we use and measure it, as well as to explain the dangers involved with its misuse. It is not intended to be a *do-it-yourself* manual or a list of projects, but just a simple, easily understandable explanation of what electricity is all about so as not to be "in the dark".

A previous technical understanding of electricity is absolutely not necessary and the book does not contain any complicated mathematics. Only one simple, easy-to-understand-and-use formula is provided—no more!

This book is intended for all persons, from curious young teens to adults, who want to have a better understanding of this important power source. Information given will also allow the layperson to begin to understand and even troubleshoot faulty devices, make simple repairs (often saving money in the process) and approach electricity with the respect it deserves.

By carefully reading and "digesting" the various chapters, one will not only attain a good basic knowledge of electricity but will be able to talk to an electrician, when necessary, knowing and understanding pretty much what he or she is explaining.

It is the author's sincere hope that everyone reading this book will learn enough to avoid being afraid of electricity, but rather to better understand it and work with it safely.

IM

I

A Brief Introduction

You may not realize it, but no one has ever actually seen electricity. It's true; not the most sophisticated research laboratory, not the most famous physicists, not even the "giants" of technology and industry, either present or in the past, have seen it, but we all certainly know what it does.

It lights our lights. It heats our homes in the winter. It cools our homes in the summer. It runs our PCs, TVs and most common household and commercial appliances. Without it doorbells, cars and airplanes would not work. Without it there would be no computers, no audio and video equipment or even mobile phones to mention just a few examples. In fact, without electricity our lives would be very, very different. We would not be much more advanced than the people using steam power in the 1700s.

So what is electricity and where does it come from? Scientists tell us that electricity is caused by the movement of tiny particles called **electrons**. These particles are so small that more than 1,000,000 of them would fit on the point of a pin with room to spare! Electrons exist in nature and are part of everything. Wood, metal, stone and even people and animals contain electrons. Although electrons are present in virtually everything that exists, the trick is to get them to move; then we have what we know as electricity.

Although it may seem impossible to move something we cannot even see, it is really quite easy. All our sources of electricity accomplish this in various ways that we will soon investigate. For now, just think of electricity as a moving flow of electrons, in many ways like a moving flow of water.

Like most flows of any liquid, a flow of water needs some sort of a conduit if we want to direct it to where we need it. Figure 1 shows a simple water system, the main purpose of which, in this example, is to run our shower in the morning. Water from a source (somewhere) flows through a closed pipe which guides it directly to our shower head.

After we use it, the remaining water then flows into the drain (and into another pipe) to be eventually disposed of. In the case of a flow of water, pipes work perfectly to direct the water to where we need it, contain it as it flows and then, after we use it, direct it to the place where it is ultimately disposed of.

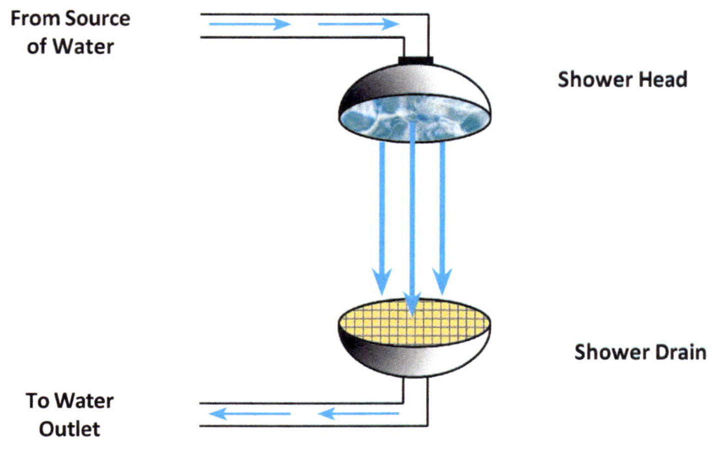

Figure 1, Simple Water-Based Shower System

Figure 2 shows the same situation, but this time using electricity to light a simple lamp. Instead of pipes, this system uses insulated wires that direct the current to the lamp where it is converted into light and then returned to its source after we use it. Both systems are quite similar in theory. Electricity flows from its source through a load (i.e., the device we want to power, such as a lamp) and then eventually back to its original source. As we continue our discussion, you will often notice the similarity of a water system to an electrical system; however, you will also see that there are important differences.

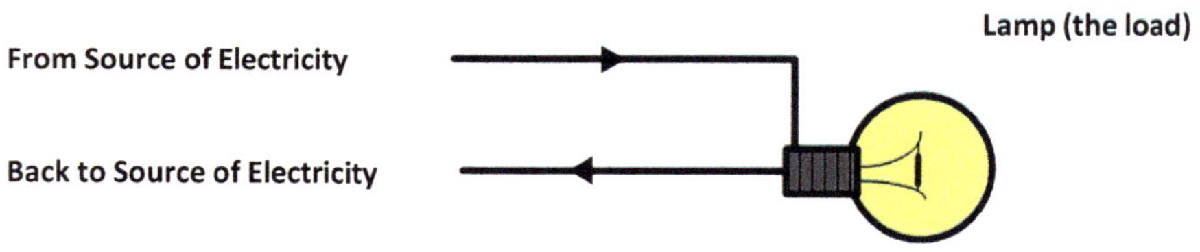

Figure 2, Simple Electrical-Based Lighting System

The pipe that carries the water can be thought of as a **conductor** of the flow of water. The wires that carry the electrical flow are also thought of as conductors of electricity. In the case of water, the walls of the pipe are strong enough to keep the water pressure inside the pipe without leaking. Electrical wires use a plastic or rubber covering to keep the electrons inside the wire as well. If the pipe leaks, we just have a puddle of water, but if the wires "leak," we can have a dangerous situation as we will soon see.

In the electrical world there are materials that can conduct electricity and materials that can block it. Those that can conduct electricity are called **conductors** because they allow it to flow. Materials that cannot conduct electricity are called **insulators** because they block it (or insulate it). Most metals are conductors while rubber, glass, and most plastics are insulators.

Figure 3 is a diagram showing a cutaway view of a common electrical wire. The centre conductor is made of metal, usually copper (a very good conductor), which is used to carry the flow of electrons, while the outer covering is made from an insulating material, usually plastic or rubber type material (very good insulators) which is used to keep the electrical flow inside the wire.

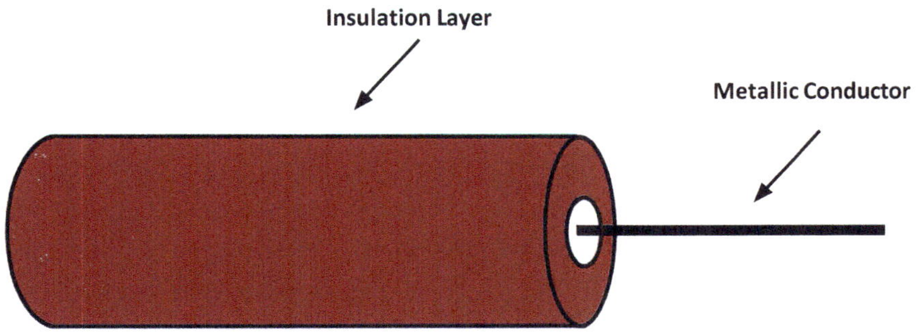

Figure 3, Cutaway View of an Electrical Wire

Referring to Figure 1, you will see that we mentioned the source of the water. In our homes, this is usually the water supply utility which we do not have much control over. This is also true of the water inlets and outlets, which we must connect to. In the electrical diagram of Figure 2 the supply is from the electricity supplier which we also cannot control but, as you will soon see, the wall sockets are what we must connect to, or no electricity will flow. At this point, Figure 1 can be thought of as a **circuit** or path for water from the input to the output. In the electrical world of Figure 2 this similar path, but this time for electricity, is also called a circuit. In both cases, however, there must be a complete path or circuit from input to output for the flow to occur. Let me explain:

In the water circuit the source of the water is the water supply utility, or possibly, a well. If the source is the local water supply, the water initially comes from a reservoir or lake or similar source.

It then goes through various pumps, pipes, filters and storage devices before finally reaching our shower system. After our shower, the water goes into the drain. The drain is connected, also through pipes, to either a public sewer or a septic tank of some sort, but it doesn't end there. The water eventually soaks into the ground, evaporates, forms clouds and then, when it rains, returns to the source. If any portion of this cycle is broken, the flow of water eventually stops. This can be caused by a broken pipe, a prolonged drought preventing rain, etc. Regardless of the cause, any break in the cycle may not stop the flow of water immediately, but if it continues long enough the flow will eventually stop.

An electrical circuit is similar in that the source is the electricity supply. There, large generators produce the electricity, then drive the flow of electricity through the wires (and through the device we want to power) and then pass it back to the source. The source can also be various types of batteries, chemical or solar, but for the flow to occur, regardless of how it is produced, there must always be a complete circuit. Unlike the water analogy, however, any break in this circuit will immediately stop the flow of electricity.

So far we have learned that in order for electricity to flow, there must be a conductor for the electrons to flow through and a complete circuit from the source, through the device we want to power (the load) and then back to the source. In the next chapter we will learn common terms concerning electricity and what they mean.

Review of Chapter 1

- What we call *electricity* is actually a moving flow of electrons.
- A continuous flow of electricity requires a complete circuit from its source and back to it.
- Electrons will only flow through a conductor and are blocked by an insulator.
- Most metals and some liquids are conductors, while most plastic, rubber, and glass items are insulators.
- An electric flow is kept inside a wire by means of an insulating coating outside the wire.

II
Some Basic Terms

Before going any further please bear with me for a tiny bit of theory that you should understand. We all know the saying that "opposites attract", which also implies that things that are similar or the same tend to repel each other. Regarding magnetism and electricity, this is true. For example, we know that common bar magnets have a so-called "north pole" and a "south pole". On most of these they are marked with a different colour, or an "N" or an "S". To prove their polarity, if you take two magnets (whose poles are clearly indicated) and place the north (or south) pole of one near the same pole of the other, the two magnets will tend to push away or repel each other. If we then turn one magnet around so that its north pole is facing the south pole of the other magnet, the two will then attract each other.

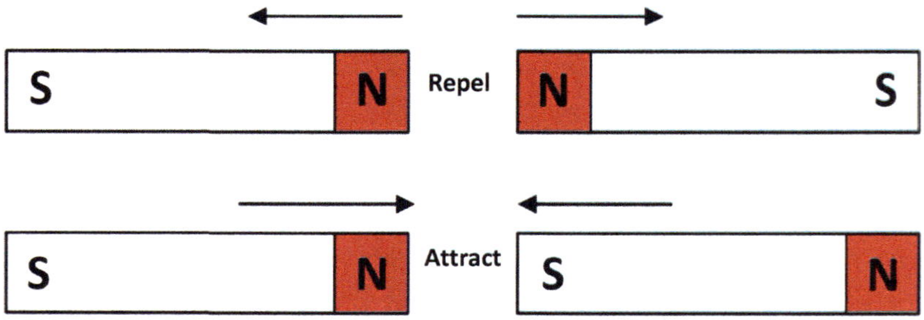

Figure 4, Opposites Attract

Atoms, of which all matter is made, have many subatomic particles, but the ones we will be concerned with are electrons. Electrons have what is called a **charge**. The exact nature of electrical charge is not important for our explanation but the one thing you should know at this point is that, in nature, a charge can be positive, negative, or neutral, meaning no charge at all. In the case of electrons this charge is always negative. Therefore, like the magnets in Figure 4, if two electrons are close to each other they will repel or push away from each other. If electrons find an area where there are positive charges (or no charge at all) they will flow toward that place. If you have a large container of electrons they will all be pushing against each other trying to escape. As we see in Chapter 1, they must have a path or complete circuit to allow them to move and carry them through the load (where they perform useful work) and then back to a positive or neutral area. When this occurs, we have a flow of electrons called electricity. We will look at why this is important a little later.

Now it is important to learn how electricity is described and measured. When electrons flow through a wire in a complete circuit, the flow is called an electrical current. The quantity of electrons flowing, which relates directly to the strength of this current, is measured in amperes (A). In the water analogy, the amount of flowing water is also called current and is usually measured in gallons. But for water or electricity to flow, some type of push is required. In the water system the push is simply called pressure. In the electrical system the push is formally called electromotive force (EMF), and the strength of this force (its more common name) is measured in volts (V). Using simple logic, it is obvious that in order for an electrical current to flow, there must be sufficient voltage to force electrons to flow.

This fact is important for the reader to understand, since the common warning "Caution High Voltage" does not by itself fully describe the potential danger. For example, everyone has experienced a small shock after rubbing their feet on a carpet on a dry winter day and then placing a finger near a metal object. Doing this will often result in a small spark, which can usually be seen (and felt). This is caused by the sudden rush of electrons from the person into the metal object. The amount of voltage needed to produce a visible spark in air is on the order of thousands of volts. But, in this example, the amount of current (measured in amperes) is extremely small. In fact, it is so tiny that there is virtually no danger to a person except for the slight sting of the spark. On the other hand, the voltage available from a household mains supply is only 230 volts (in the UK and most of Europe), but the amount of current available is thousands of times greater and can easily cause severe pain or even death. As a result, any warning notice should never be ignored.

To get an idea of the strength of an electrical current consider the fact that a simple AA battery only has 1.5 volts. It can however push enough current (about 1/3 of an ampere) to light a small torch bulb at a brightness of around 1/2 watt. The household mains supply in the UK, on the other hand, can provide many amperes of current, enough to easily run a refrigerator, a vacuum cleaner, and a 1000-watt heater. With regard to electrical danger to humans, it has been shown that only 20-millionths of an ampere flowing directly through the human heart will cause problems. On the other hand, while 200 to 300 amperes from a 12-volt automobile battery is needed to start a car, the voltage is usually too low to force dangerous current through intact skin and reach the heart.

Therefore, accidentally touching the battery terminals is not necessarily dangerous. Clearly, when talking about electricity, both voltage and the strength of the current (amperes) must always be considered together. To quote an old song, *"you can't have one without the other."*

In all systems there is always some **resistance** to the flow. This can be a shower head with small holes that produce a strong stream of water from a larger-diameter pipe in the water circuit or a lamp with a small filament that must use electrons to produce light in an electrical circuit. Resistance in the electrical world is anything that opposes the flow of electrons through a circuit. It is measured in **ohms (Ω)**. We will now see how these are related to each other.

There is a simple relationship, developed many years ago, that describes many electrical circuit parameters. This relationship is expressed as a simple formula called **Ohm's Law**. Figure 5 shows this in basic mathematical terms as well as in a commonly used simple diagram.

But before the math in the following paragraphs "turns you off" please note that this will be the absolute extent of the mathematics used in this book. It is all that is really needed at this point, but the calculations are important and will be helpful to you in the future toward your understanding of electricity.

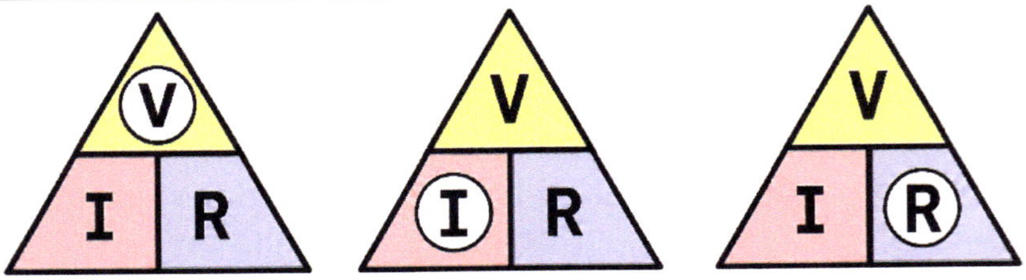

Volts = Amperes × Ohms

Watts = Amperes × Volts

Amperes = Volts / Ohms

Ohms = Volts / Amperes

Figure 5, Ohm's Law (Voltage = V, Current = I, Resistance = R)

The "I" used for amperes signifies "Intensity". The "R" used for ohms stands for "Resistance"

Cover the quantity you want to find. The remaining quantities show the calculation required.

The one parameter not illustrated in Figure 5 is electrical power, measured **watts**. Since electrical power requires current (measured in amperes) and voltage (as we have seen), watts are simply equal to voltage times current. That is why our 1.5-volt battery must push approximately 0.33 amperes of current through the circuit to light our 1/2-watt lamp (1.5 × 0.33).

When indicating voltage, current or resistance, metric abbreviations are always used to indicate values. Refer to the chart for examples.

PARAMETER	METRIC	DESCRIPTION	EXAMPLE
Volts	kV	Kilovolts	1 kV = 1000 volts
	mV	Millivolts	1 mV = 0.001 volt
	μV	Microvolts	1 μV = 0.000001 volt
Amperes	mA	Milliamperes	1 mA = 0.001 ampere
	μA	Microamperes	1 μA = 0.000001 ampere
Ohms	kΩ	Kilohms	1 kΩ = 1000 ohms
	MΩ	Megohms	1 MΩ = 1,000,000 ohms
Watts	μW	Microwatts	1 μW = 0.000001 watt
	kW	Kilowatts	1 kW = 1000 watts
	MW	Megawatts	1MW = 1,000,000 watts

By understanding the relationships between the parameters, you can determine the power needs of appliances you have. For example, if you have a computer that draws 2.5 amperes from a 12-volt source, you can assume that it is consuming approximately 30 watts (W = V x I). Similarly, if you have a 1000-watt heater that you use on a 230-volt AC (alternating current) supply, it will draw approximately 4.35 amperes (I = W/V). In this manner, you can estimate how much electric power any appliance draws and, by multiplying its usage with the cost for electricity in your area, you can easily calculate how much it costs to operate. Note that these are only approximate values and will depend on the specific device you have.

If you do not know the internal resistance of a device, you can determine it (in ohms) by dividing the voltage it requires by the current it will draw. Then, when you know the resistance of a device, you can obtain the other parameters. For example, the computer we discussed earlier, which requires 2.5 amperes of current at 12 volts, has an internal resistance of approximately 4.8 ohms (R = V/I), while the 1000-watt heater, which requires 4.35 amperes of current at 230 volts, has an internal resistance of approximately 53 ohms (R = V/I).

The calculations described by Ohm's Law are accurate for **DC** voltages and currents. **AC** voltages and currents are similar, but additional factors can affect these calculations. Those factors are beyond the scope of this discussion. For AC, these simple calculations are accurate enough to get approximate results. Note: DC and AC are explained later.

Figure 6, Typical Electrical Requirements of Common Household Appliances

Review of Chapter 2

- The flow of electrons in a circuit is called *current* and is measured in *amperes*. The pressure available to move or push these electrons is measured in *volts*. The resistance to an electrical current flow is measured in *ohms*.

- To have an operating electrical circuit, you must have both voltage and current. You can't have one without the other.

- *High voltage with extremely low current will not necessarily be dangerous, just as high current without significant voltage will also not necessarily be dangerous, but never ignore any warning or caution indicator when using electricity since you may not know all the actual parameters. You must know all the relevant parameters to determine the real level of danger.*

- The power available in an electric circuit is measured in *watts*, which is a function of volts and amperes.

- The relationship between volts, amperes and ohms is described by *Ohm's Law*. Watts are simply volts multiplied by amperes in a particular circuit.

- Metric terms such as *milli* (one-thousandth), *micro* (one-millionth), *k* (thousand) and *M* (million) are used to indicate lesser and greater values.

III

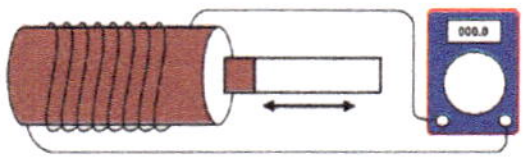

Sources of Electricity

Now that you have some idea of what electricity is and how it is measured, the next step is to see how it is produced. In Chapter 1 we learned that electricity is a flow of electrons and that these electrons have a negative charge which means that they will tend to repel each other. If we were to have a container of these electrons, they all would have a negative charge and would be trying to push against each other to escape from the container. One such container is a common battery. As a point of information, although the term "battery" literally means a *group* of things, the term is commonly used to describe devices that produce DC electricity. Technically, however, a "battery," such as the common AA or AAA battery, is made up of just one simple cell. A 9-volt battery, on the other hand, is a true battery since it is usually composed of six 1.5-volt cells connected internally as shown in Figure 7, while an automobile battery is composed of six 2-volt cells. Common usage, as we mentioned, has labeled anything from a tiny wristwatch cell to a 21-kilogram automotive device a "battery". So, keeping with convention, we will call any such device a battery as well.

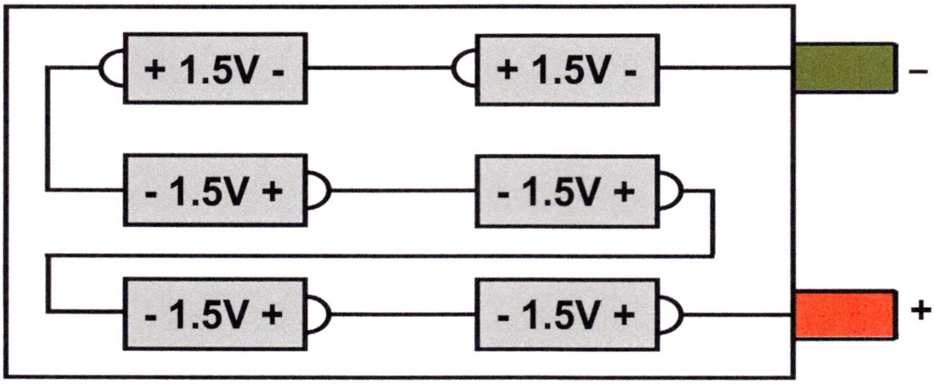

Figure 7, Internal View of a 9-Volt Battery

Since a simple battery can be considered a container of electrons you might ask how the battery creates electricity. This is produced within the battery through chemical action. In most batteries there are materials that, in combination with each other, produce electrical energy.

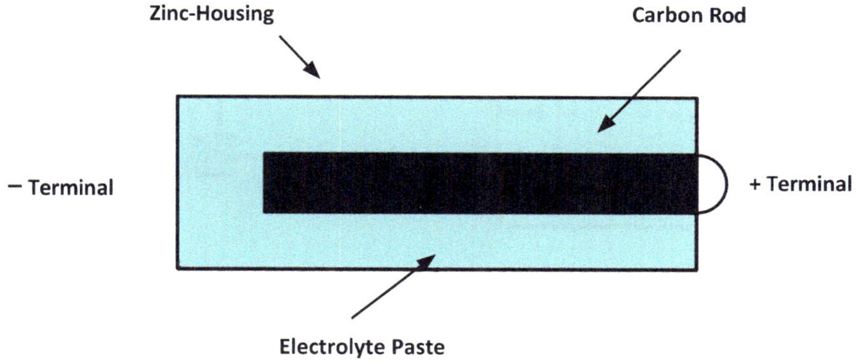

Figure 8, Common Battery

Figure 8 is a cutaway diagram of a typical AA battery which, in this case, is an example of an old-fashioned so-called **zinc-carbon battery**; but this figure shows simply what is inside. There is a carbon rod, a zinc metal housing, and a electrolyte paste that surrounds them. Not shown is an outside covering which is usually made of a non-conducting material. Due to the chemical reactions between the internal materials, this combination is what creates the electrical potential of 1.5 volts we are familiar with. The electrolyte paste then acts, in conjunction with the other materials, to enable electron flow. Modern batteries use different materials for better reliability and longer life, and although the voltages are somewhat different, the result is essentially the same. When the chemical filler is "used up" the battery dies. How and why this action produces electrons is not necessary for our discussion so at this point let us just accept the fact that chemical action in the battery creates an electrical potential that allows electrons to flow.

There are many types of batteries in use today. They come in sizes ranging from tiny hearing-aid and wristwatch types to the huge versions powering cars and lorries, not to mention numerous other applications. The zinc-carbon type we just described was the original portable **dry cell** battery (although the internal electrolyte paste was not really dry, but a sort of moist paste). In the process of producing electrical current, the zinc outer housing slowly dissolved over time until the electrolyte paste eventually leaked out or was used up, limiting the amount of current that could be produced. Those batteries also often created a mess or corroded contacts when they leaked. They were replaced by **alkaline batteries** which used a different combination of materials and lasted much longer. Today there are many types using all sorts of materials. Manufacturers' ultimate goal is to make batteries that will last as long as possible.

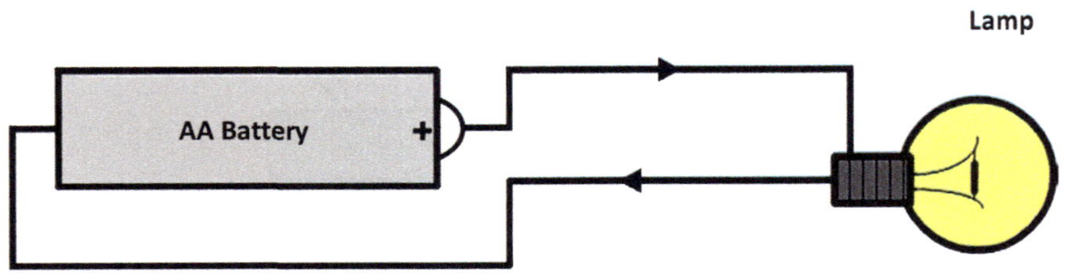

Figure 9, Simple Electrical Circuit

As I am sure you are aware, most batteries have a terminal called **positive** "+" and a second one called **negative** "–". Before any connections are made, the electrons are trapped and pushing against each other inside the battery but cannot escape. If we now connect a light bulb in the simple circuit shown in Figure 9, there will be a complete path between the + and – terminals through the bulb. Now electrons can flow, and the lamp lights.

Note that the arrows in the drawing show that the electrons are flowing from + to - even though we know that electrons repel each other and therefore should be going in the opposite direction. Unfortunately, this direction of flow was accepted by scientific convention before electrons were discovered. In reality, negatively charged electrons move toward the positive terminal (which is in the opposite direction). It is confusing, but once a scientific convention is made, it is difficult to correct so we are stuck with it and will continue to use it even though it is not accurate.

Some of the other common batteries we might encounter are the so-called **wet cells** used in our cars, the **gel cells** used in emergency lights, and backup UPS* systems used for computers as well as in various marine applications. Unlike dry cells these batteries use metal plates for the electrodes and a liquid or gel such as sulfuric acid to produce electric current. Some of these also have the unique ability to be re-charged once the internal chemical process is depleted. There are even solar batteries which convert sunlight into electricity using unique semiconductor materials. The result of all of these is the ability to produce a continuous flow of electrons to operate the various devices our society uses.

So far the sources of electricity that we have looked at come from batteries, be they chemical (of various compositions) or solar. What they all have in common is that the electricity produced is called DC or **direct current**. This means that the flow of electrons is always in one direction, from + to – (really from – to + for the reason we have explained).

**UPS (Uninterruptible Power Supply)*

Obviously the electricity that comes to our homes does not come from batteries but from the electricity supplier. Batteries are not typically used since the high electrical demands of towns and cities are much greater than what they can provide. The most common source used by the electricity supplier is a device called a generator. This is a huge electro-mechanical device capable of producing tens of thousands of watts of power per device, and megawatts of power per power station. To help understand how such a device works we will perform an experiment. This will require a simple tool (which we will discuss in depth in the next chapter), a permanent bar magnet of any type and some insulated wire. But first let us look at the common tool we just mentioned that most people should own. Figure 10 shows an easily obtainable **digital multimeter**. It is a device used to measure voltage, current and electrical resistance (ohms). Such a device is quite inexpensive, and you can easily obtain one (and really should) from a local DIY store or online. You should shop around a bit but not have to pay more than £10 or so (for a new one) as you only need the simplest one available for some of the requirements in this book and for future electrical projects.

Figure 10, Typical Digital Multimeter

The multimeter shown in Figure 10 is similar to most you will find in this price range. We will look at how to use this instrument in the next chapter. But for now, you can see that this unit has a large rotary dial that allows you to choose what parameter you wish to measure and two probes (usually red and black) to connect to what you are measuring. For our present investigation, however, we will set the meter to the "200m" (200 milliamperes) range which is the most sensitive current range. On the specific multimeter you obtain, the notations may be somewhat different, but you should choose the most sensitive current range on yours. We will investigate the other ranges and functions in the next chapter.

For this demonstration you will also need some insulated wire (such as common **bell wire** from a local hardware store), a simple bar magnet and a cardboard tube such as the ones kitchen roll or toilet paper usually come on. Referring to Figure 11, wind about 10 to 20 turns of wire around the tube. The type of wire and tube is not particularly important, just make the turns close enough to touch each other. You can then use any sort of tape to keep the turns in place when you are finished. To begin, plug the black multimeter probe into the **COM** jack and the red multimeter probe into the **Volt/Ω/mA** jack of the multimeter. Next connect the ends of the wire coil to the probes. If the wire does not have the centre metallic conductor exposed at each end remove or scrape off a little of the insulation to allow the metal tips of the probes to contact the metal conductor inside the wire. Now quickly move the magnet into and out of the cardboard tube (and the coil) and you will see the numbers on the meter change to both positive and negative readings. It is not important just what these numbers mean for now, but what is important is to understand that the **mechanical energy** of the moving magnet is being converted into electrical energy and the multimeter is showing the result.

When the magnet moves into the coil, electrons begin to flow in the coil and they flow in one direction. When the magnet is moved out of the coil, electrons also flow, but now they move in the opposite direction. You might see positive numbers in one direction and negative numbers in the opposite direction on the multimeter. Since the current is changing direction with the magnet's movements, this type of current is called **alternating current** or AC. We are not able to produce much AC current with our simple setup; however in electrical power plants gigantic coils of wire are moved in close proximity to huge magnets by various means and very large quantities of electricity are produced.

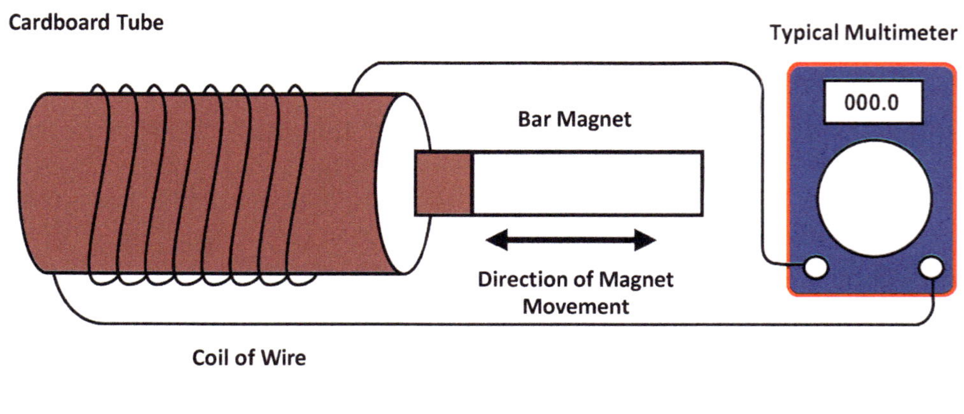

Figure 11, Explanation of Generator Operation

As in the case of the battery circuits we have been discussing, a complete circuit (or path) to and from the source is still needed, and that is why the mains leads and outlets in our homes and on our various appliances all have a minimum of two wires. In addition, since the direction of flow of the electrons comprising an AC current is constantly changing, the "+" and "–" designations are not used, but a complete circuit is always needed.

Review of Chapter 3

- As a point of information, although the term *battery* literally means a group of things, it is commonly used to describe practically any device (i.e., even a single device) that produces DC electricity.

- Current flowing in only one direction is called *DC* or *direct current.*

- The direction of flow of electrons from a DC source is really from "−" to "+", but scientific convention shows it from "+" to "−".

- Electricity suppliers use huge machines called *generators* to produce large amounts of electricity for use in homes, villages and cities.

- A digital multimeter is an instrument used to measure voltage, current and resistance.

- When a magnet moves into and out of a coil of wire, electrons begin to flow. The direction of flow of this current changes as the magnet moves into and out of the coil. This type of current is called *AC* or *alternating current.*

IV
Tools of the Trade

In Chapter 3 we used the digital multimeter (multimeter) for a simple demonstration to understand the difference between DC and AC voltage and current. Now we will examine the multimeter in more detail, as it is one of the most common and useful tools you will require when working with electricity. You should obtain one now if you have not already done so.

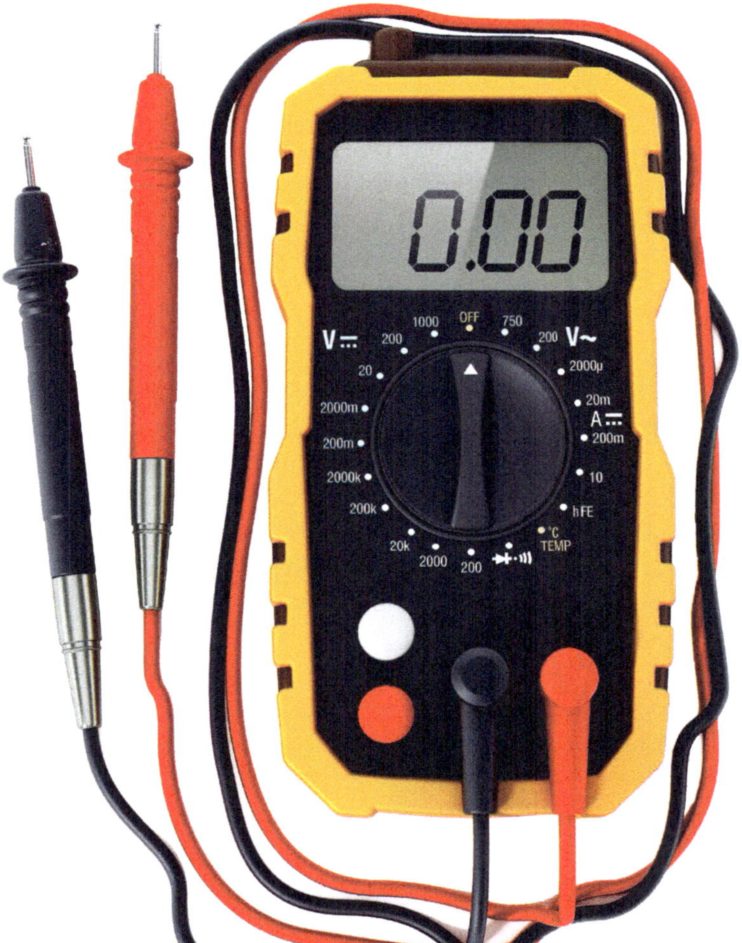

Figure 12, Typical Digital Multimeter

If you refer to Figure 12, you will see that the multimeter has a central selector dial, two probe connectors and a couple of buttons. The multimeter you have may appear somewhat different, but the functions will still be basically the same. All multimeters come with two probes used to connect to the portion of the circuit you wish to measure. In the next steps we will use the multimeter shown in Figure 12 as it is one of the most common types. If yours is not exactly the same, do not worry. It will be close enough to do what is necessary.

To use the multimeter proceed as follows: first plug the probes into the connectors on the multimeter. For our first example we will measure DC voltage. Plug the black probe into the common connector, usually labeled COM for most voltage and resistance readings. Next plug the red probe into the connector labeled V/Ω/mA or similar. The third connector, if present, is reserved for high current readings which we do not need at present. Now turn on the multimeter. On some units there will be a power switch; on others the power will turn on when you move the central selector dial to the value you wish to measure. Some units may also have a "Hold" button which you can ignore for now.

Obtain a common AA or AAA battery. Turn the central dial to the DC voltage position indicating 2 volts or more (in this example 20 volts). Others may indicate "2V". This position will display voltages from 0 to 2 volts. Now turn on the multimeter if it is not already on. At this point the display should read "1.000". If for some reason "Hold" is displayed (on multimeters with Hold buttons) push the Hold button to allow a reading. Next touch the tip of the red probe to the + side of the battery and the tip of the black probe to the - side at the same time. The meter should now read 1.50 volts or thereabout. If you reverse the probes, red to – and black to +, the meter will read -1.50 volts.

So you can read voltage and determine its polarity as well. Note that when making DC voltage readings (especially unknown ones) it is always important to start with the highest voltage range of the multimeter to avoid damaging it, and then gradually reduce the setting one step at a time until the reading is displayed with the resolution you desire. You will also see other labeled DC voltage ranges available which can be used to measure DC voltages, often up to 500 volts. Note that all ranges displayed on the dial are the maximum that can be measured at that setting.

There is another voltage range, and this is usually labeled V≈ or VAC. This is for measuring AC voltages. When using the meter to measure AC, as in the instructions for DC, it is best to start with the highest AC range and then gradually reduce the setting, one step at a time, until the reading is displayed with the resolution you desire. Never use the DC ranges to measure AC voltages or you may damage the multimeter. Similarly, never use the AC ranges to measure DC voltages or you will either get no reading at all or erroneous values.

When you wish to measure AC voltages, such as at the outlets in your home, you must use the AC ranges. The probes must be connected to each side of the outlet. When doing this, be extremely careful not to touch any exposed metal portion of the probes as you insert them into the wall socket, since doing so could result in a severe electric shock. As long as you touch only the insulated portion of the probes, however, you will be fine. Our sample multimeter has two AC ranges, 200 volts and 500 volts, although some meters intended for use in the UK may have higher ranges because the nominal mains voltage is 230 volts. As with the DC ranges, first set the multimeter to the highest AC range, connect the probes, and then reduce the setting if necessary. You should see a reading of around 230 V in the UK and most of Europe. If you accidentally use the DC ranges instead of the AC ranges for AC measurements, you may damage the meter, so be very careful.

The final measurement range we will discuss for now is the resistance ranges sometimes designated by the omega symbol (Ω). You will see a series of values from 200 Ω to 200 MΩ in this range. These values refer to ohms, with 200 meaning 200 ohms and 200 MΩ meaning 200 megohms (200 million ohms) maximum. There is also a range in this area on some multimeters labeled with a beeper symbol))) or "20" ohms or something similar. When you set the multimeter to this position and touch the tips of the two probes to each other, you will hear a beeping sound. This range is used to check the presence of **continuity** (an extremely low resistance path that you would get from a good connection) between the two probes and will be very useful as you will soon see. When using the resistance ranges, make sure you never connect the multimeter to a **live** circuit, either DC or AC, or you can easily damage it. *Make absolutely sure that all power connected to whatever you are measuring is off.*

In addition to the multimeter, several other tools are also commonly used when working with electricity. Figure 13 shows **wire strippers** which are used to remove the insulation from electrical wires without damaging the metal conductor. Figure 14 shows **long-nose pliers**, **diagonal cutters** and **slip-joint pliers.** All of these are used to bend, cut and manipulate wires and various related accessories. While they are made of metal, you will note that they all have insulated handles.

The tools mentioned are relatively inexpensive and easy to obtain. If you are planning to do any electrical work, it is advisable to obtain a good quality version of each. A set of common flat-blade and Phillips or cross-head screwdrivers (specifically with insulated handles) is also required as is a roll of common black electrical tape. When purchasing such tools, always try to find ones that are of decent quality. Very inexpensive tools will cause problems later on.

Figure 13,
Common Wire Strippers

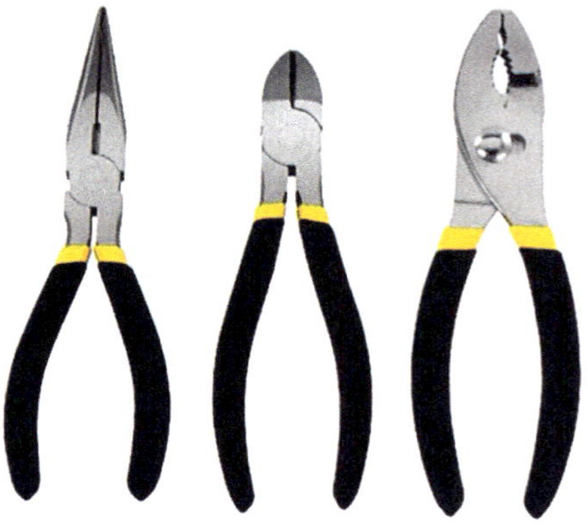

Figure 14,
Typical Electrical Tools

Note that while all the tools indicated above have insulated handles, no work should ever be carried out on a powered or live circuit where AC or even DC voltages are present. As mentioned in Chapter 2, when sufficient voltage is present to force enough current through your body, the result can be disastrous.

Review of Chapter 4

- A *digital multimeter* or *multimeter* is an instrument used to measure voltage, current and resistance.

- When using a multimeter to measure voltage or current, always start from the highest range and gradually decrease the range until you get a reading.

- When measuring DC voltage, use only the DC voltage range. When measuring AC voltages, only use the AC voltage range. Do not use the wrong range or you may damage the multimeter.

- *Never try to use the resistance range in a live circuit. Always make sure that all power is off before trying to use this range.*

- *Do not touch any exposed metal portion of a multimeter probe, or you may receive a serious electrical shock. Handling the insulating plastic probe body, however, is normally fine.*

- *When using any of the tools described, be sure that all power to the circuit you are working on is turned OFF. Never use any tool on a "live" circuit whether it has insulated handles or not!*

V
Basic Electrical Components

It is time to look at some of the typical basic components we will come into contact with when dealing with electricity. The first and one of the most common is the simple **switch**. A switch is used to turn power to an electrical circuit on and off, whether to start our car, turn on our television or mobile phone or to simply control the lights in our home or in a factory. As we learned in Chapter 1, there must always be a complete circuit from and back to the source for anything to operate from electric power. If the circuit is broken at any point, the electrons will not flow, and stopping the flow is exactly what a switch does. Figure 15 shows several types of switches. Although they may look very different from each other at first glance, they all do the same thing. They either complete a circuit or break a circuit, allowing or preventing the flow of electrons. If you can obtain a switch, whether it is one designed for your home or for some other purpose, look at it carefully. In virtually all cases you will see metal terminals of some sort used to connect wires to the switch. Figure 16 shows a common household wall switch. This particular switch has three terminals. Two are used for the live connections and the third is the earth terminal..

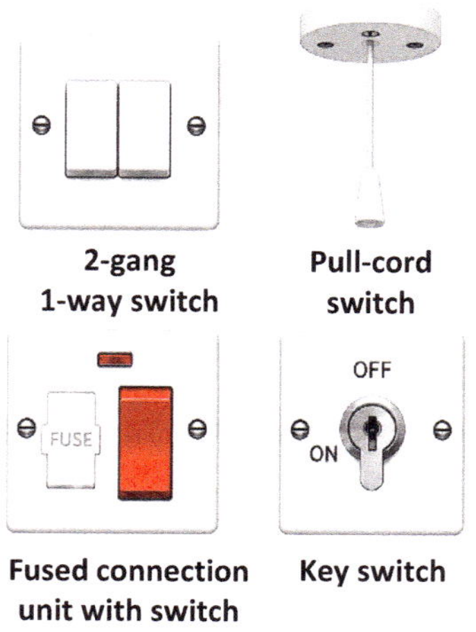

Figure 15, Several Electric Switches

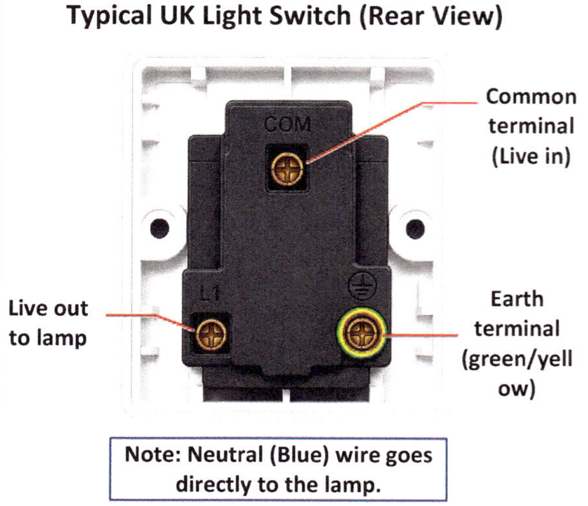

Figure 16, Common Switch (UK)

To demonstrate how this switch works, take your multimeter, set it to the lowest resistance range (the one where the probes sound a beep when touched to each other), and put one probe on one connecting screw and the other probe on the other connecting screw and then activate the switch. In one position of the switch (the "On" position), you will hear the beep, and in the other position (the "Off" position), there will be no beep. If your multimeter does not have an audible beep setting, just use the low resistance range which should read 1.000, depending on the setting of the switch. This method can be used to check the operation of virtually any switch. You can ignore the earth terminal for now, as it is used for protection on an AC wall switch, a topic we will look at later.

Figure 17, Barrel, Car Battery, and Crimp-Type Connectors

Connections to DC power sources usually range from clamps or clips to small connectors. Figure 17 shows some common ones, but these are only a few types. Many more will be found from the small round **barrel plug** used on many portable devices to the heavy clamps used on car batteries. Since the voltages and currents usually employed with these connectors are fairly low and not normally dangerous to human beings the amount of insulation used is not excessive. In some cases currents can be high, however, such as with car battery clamps, so while some types may be large in order to handle high currents, the voltages are normally low enough so there is usually little danger to people accidentally coming into contact with exposed conductive parts. *Although high currents may not necessarily be dangerous from an electrical shock point of view, they can sometimes cause electrical wires to become dangerously hot to the point where they can cause a serious burn, so be careful.* Nevertheless, when working with any type of connectors that have exposed metal connection terminals and carry high voltages or currents it is important to be careful and to avoid physical contact with these terminals.

For AC power connections the situation is different. In this case, particularly when connecting to the mains supply, voltages are high enough that enough current can flow and cause serious harm to people coming into contact with any conductor, and as a result the construction is different. Figure 18 shows a common mains socket and plug. You will note that the construction is more rugged and that all metal conducting points are made so that it is very difficult for a person to physically come into contact with them when they are in use.

Figure 18, Typical UK 3-Pin Mains Socket and Plug (BS 1363)
230 V AC, 50 Hz Supply

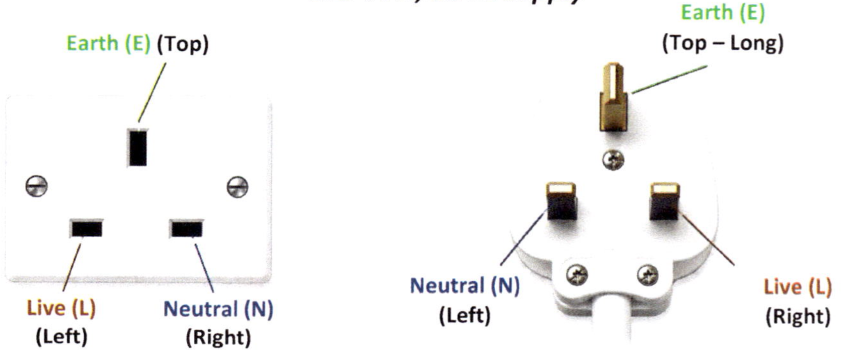

The earth pin (Green/Yellow wire) connects to the top longer pin.
It provides safety by directing fault current to earth if a fault occurs.

You will also note that, like the wall switch, the wall socket has several screws, including one green screw. The socket shown is composed of two separate sections. As mentioned, the earth terminal is used for protection on household plugs and sockets, a topic we will now explain.

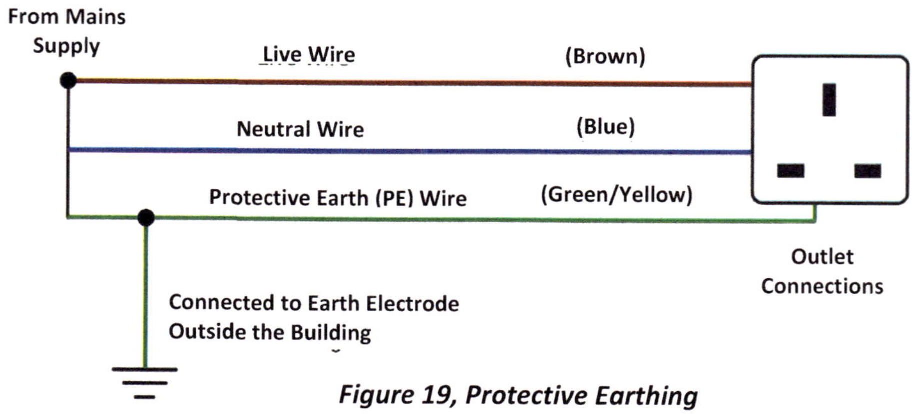

Figure 19, Protective Earthing

Figure 19 shows a diagram of a typical UK household electrical supply connection from the electricity supplier to the home and Figure 20 shows what a typical mains distribution cable looks like. You will see that the path from, and back to the supply, is composed of brown and blue wires. The green/yellow (or sometimes bare copper) wire is used for safety purposes and is not normally intended to conduct electricity. As shown in Figure 19, the brown wire is connected to the live supply. The blue wire is connected to the neutral point and is also ultimately connected to the earth wire somewhere outside the home. Sometimes a bare copper earth conductor will be used instead of a green/yellow insulated one.

Figure 20, Typical UK Mains Cable

The green/yellow (or bare copper) wire is provided to connect to an earth connection, which is often formed by an earth electrode, typically a copper rod driven into the ground outside the building. Sometimes it is even connected to a long length of buried metal pipe such as a water main feed pipe. The reason for this wire is safety. In any electrical appliance with a metal housing, control panel or metallic decoration, the metal is usually connected to an earth wire inside the appliance. The earth wire inside the appliance is then connected to the round pin in the appliance's mains lead. This is then eventually also connected to the earth (and the blue neutral wire), but at the point in the home where the wires from the electricity supplier are located. This means that if for any reason anything inside the appliance becomes damaged, a person coming into contact with any metal portion of the appliance will be far less likely to receive a dangerous electrical shock. In fact, the only live wire in the appliance is the brown **wire,** and if it accidentally touches any portion of the appliance, an internal fuse will blow cutting off the current flow. Therefore, the green or bare copper wire is often called a **safety earth wire**.

Most items intended for use with the mains supply are provided with an earth terminal or a green/yellow earth wire for this purpose, and in many cases, this is required by regulations. An earth connection is used not only in homes but virtually everywhere that electrical power is distributed. The only items that do not use separate earth wires are those where there is no possibility of a person coming into contact with any exposed live electrical conductor.

Small appliances such as lamps or food mixers (for example) with no exposed metal parts are often designed so there is no possibility of a person coming into contact with any live electrical conductor. Many of these appliances use 2-core mains leads and do not require a separate earth connection because the outer housing is made entirely of insulating material. In the UK, standard mains plugs and sockets are designed with built-in safety features, insulated pins, and earth connections where required. Figure 21 shows a typical UK 3-pin mains plug and socket arrangement.

Figure 21, Typical UK 3-Pin Plug and Socket

In Chapter 7 we will see some examples of how these are used.

Review of Chapter 5

- An electrical switch is used to make and break an electrical circuit to control when it is required for operation.

- DC power connectors are usually not as rugged as AC ones, since DC voltages are usually low enough that there is little danger that enough current can flow to harm a human being.

- *High DC currents can sometimes heat connecting wires to the point where they get very hot. So when working with high DC currents, be aware of this.*

- AC power connectors, particularly those used for mains applications, are much more rugged and have no exposed metal portions that a person can accidentally touch and receive a dangerous shock from.

- In the UK, the live wire is typically brown and the neutral wire is typically blue. The neutral is connected to earth at the supply point. The earth wire is green and yellow (striped) and provides a protective connection. In some cases, the earth conductor may be bare copper, and it is connected to earth via an earth electrode, typically a copper rod driven into the ground outside the building.

VI

Common Electrical Symbols and Diagrams

Electricians and electrical technicians normally describe electrical circuits with special diagrams that show the connections between all portions of a circuit. We saw several pictorial diagrams in Chapters 1 and 2 but in the field of electricity these connections are shown by a **schematic diagram** (often simply referred to as **a schematic**). These diagrams use standardized symbols to represent components such as batteries, switches, lamps, and other devices rather than actual pictures. These symbols are universally accepted and understood, and make the life of an electrician much easier. As an example, Figure 9 in Chapter 3 shows a very basic flashlight hookup. Figure 22 shows the equivalent electrical schematic diagram of the same circuit with the exception that a switch has been added to allow the light to be turned on and off.

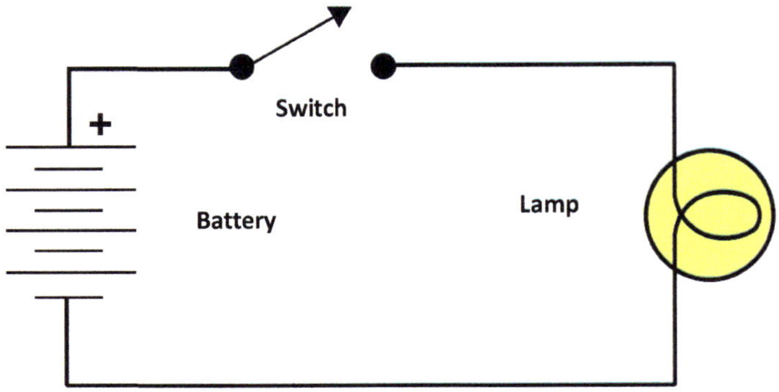

Figure 22, Schematic Diagram of Figure 9

As you can see from the schematic, the battery is represented by long and short lines which symbolize + and - plates derived from an old-fashioned wet cell. The switch is shown as an arrow that can be in the "off" position as shown in the schematic (no connection to the light bulb) or in the down position when it is conducting current and is "on". The lamp itself is simply shown as a coiled filament inside a glass bulb. This is a simple schematic diagram, but as you will soon see, schematic diagrams can become quite complex.

Figure 23 shows some of the more common symbols used in electrical schematic diagrams. You can see that in many ways they represent the action or function of the component they are describing. Keep in mind that the symbols shown here are only a few. There are many more that you will come into contact with, but with some research it will be easy to understand what they represent. Sometimes a combination of several symbols will be contained within a box, forming a more complex symbol when they interact to form a more complex component.

In the next chapter we will see some typical examples of these symbols used in schematic diagrams of equipment that is commonly used.

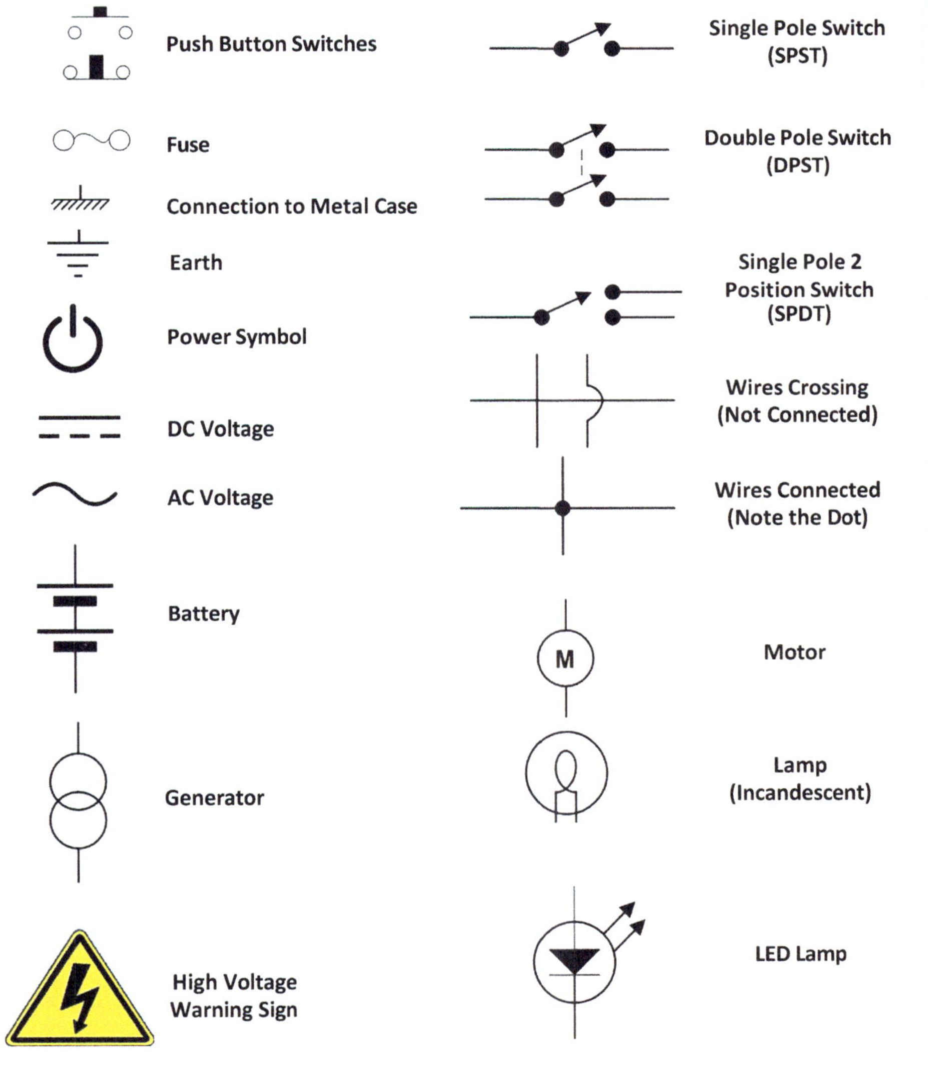

Figure 23, Basic Electrical Schematic Symbols

Review of Chapter 6

- Electrical circuits and connections are represented by drawings called *schematic diagrams,* or sometimes simply *schematics*.
- Circuit electrical components are usually represented by universally understood symbols.
- Often these symbols are designed to explain the function of the component they represent.

VII
Reading and Understanding Schematic Diagrams

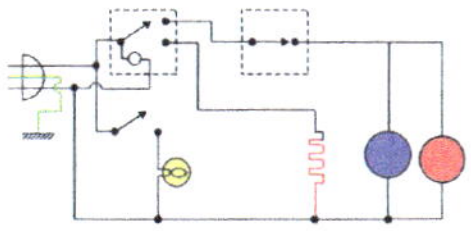

We will now look at some basic circuits you may find around your home to see how the schematic symbols which we saw in the previous chapter are used, and also to gain a better understanding of an electrical schematic diagram.

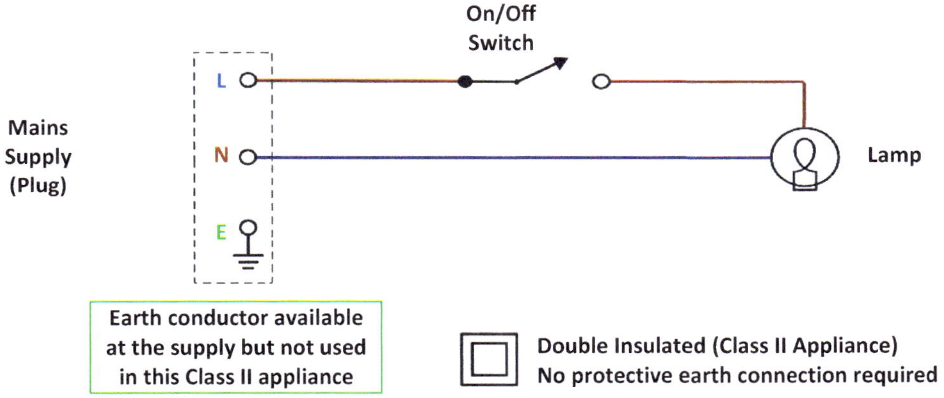

Figure 24, Common Table Lamp Schematic Diagram (Double Insulated)

Figure 24 is the schematic diagram of a common, simple table lamp. Many small lamps do not have an earth wire but use two-core flexible cable for connections, so if the third earth wire is not present this is normal in this case. These types of lamps do not have any exposed metal parts that a person can come into contact with. If you replaced the mains plug symbol in the schematic with the connections to a normal household mains supply, you would then have the schematic diagram of a common switch-controlled lighting circuit in the home.

Moving a step further, you are probably familiar with a situation where two switches control a single light fitting. Figure 25 shows the schematic diagram of this common type of circuit. The single-position switch has now been replaced with two two-way switches. As shown by the schematic symbols, each switch has a movable common contact that can connect to either one of two separate contacts, depending on how the switch is operated. If you follow the circuit carefully, you will see that if both switches are connected so that there is a continuous path from the live wire to the light fitting, the lamp will light. If either switch is operated, the lamp will go off since the circuit will be broken. When the other switch is then operated, the lamp will go on again. The connecting wires between the two switches are part of the normal circuit wiring. In UK practice, the live wire is typically brown and the neutral wire is blue. A separate earth (green/yellow) wire is used for safety and is connected to any exposed metal parts of the switches and the light fitting. This ensures that anyone coming into contact with metal parts of the installation will be far less likely to receive a dangerous electrical shock.

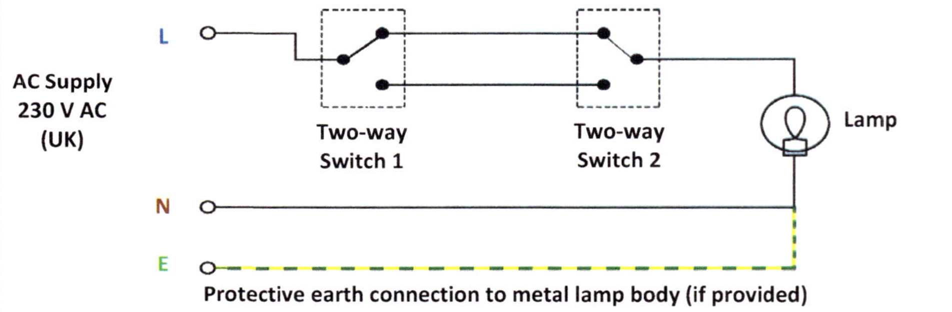

Figure 25, Two-Switch (Two-Way) Light Control Schematic

Figure 26 is a schematic diagram of a simple vacuum cleaner. The circuit is similar to the lamp circuit of Figure 24, but here the on/off switch controls two items: the power indicator lamp and the motor. Also present is a protective fuse that will open the circuit if there is an overload caused by a clogged motor, an internal short circuit, or any other condition that would cause the vacuum cleaner to draw excessive current.

This fuse may be **resettable** in that after a period of time (or after you first shut off the system and then turn it back on), it will allow the system to operate again, or it will continue to turn off the system until the fault is corrected. Regardless, at this point the circuit should be easily understandable.

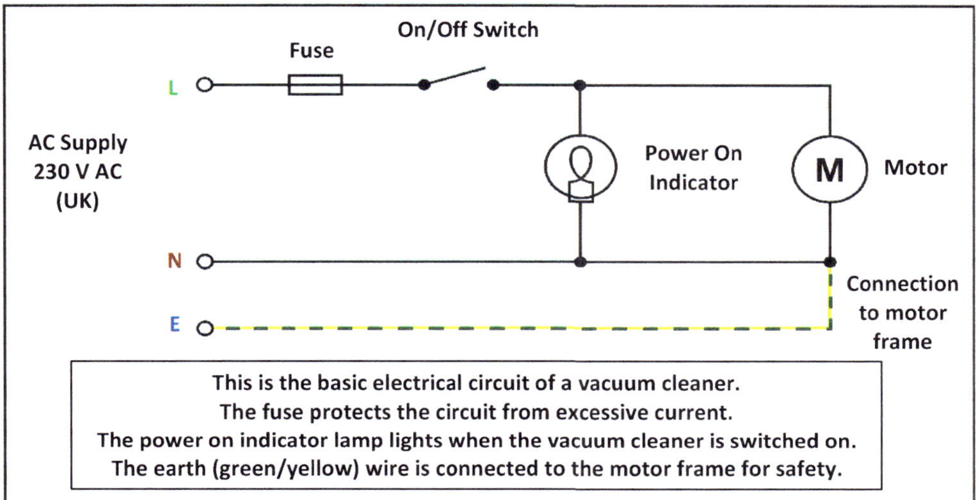

Figure 26, Simple Vacuum Cleaner Schematic Diagram

Circuits such as these are common, and you will find that by replacing components such as lamps, switches, and motors, the basic circuit remains essentially the same, even though the appliance may differ. More complex devices will, of course, result in more complex schematic diagrams.

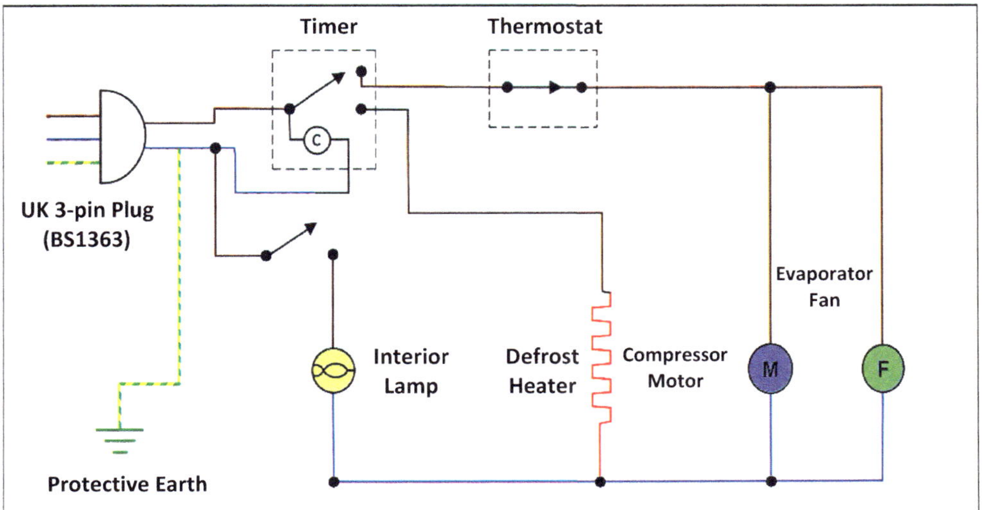

Figure 27, Basic Frost-Free Refrigerator Schematic Diagram

Our last example, Figure 27, shows a basic self-defrosting refrigerator. As you can see from the schematic diagram, this appliance is somewhat more complex, but it should still be easy to understand. The operation is described below. You should be able to follow this explanation by referring to the diagram.

1. When the refrigerator is connected to the mains supply, the timer switch (with its internal motor) and the thermostat will be in the positions shown in the diagram.

2. Since the timer motor is connected to the mains supply, it will begin timing. Note that the timer motor is usually contained within the timer assembly.

3. Since both the timer switch and the thermostat form a complete circuit from and back to the mains supply, through the compressor motor and blower, the refrigerator will begin to cool.

4. When the refrigerator reaches the temperature set by the user on the thermostat, the thermostat will open the circuit and turn off the compressor motor and blower.

5. When the temperature rises above the set point, the thermostat will close the circuit again and the cooling cycle will repeat.

6. After a period of time, ice will tend to build up in the freezer section. At a preset time determined by the manufacturer, the timer will operate its internal switch (a two-position switch). This will turn off the compressor motor and blower by opening their circuit and will switch on the heater.

7. The heater will operate for a predetermined time (also set by the manufacturer), melting the accumulated ice. It will then automatically return to its original position and the cooling/defrost cycle will repeat.

8. You will also notice that whenever the refrigerator door is opened, the interior light will come on, as it is connected to the AC supply through the door switch.

This is a simplified schematic diagram of a basic refrigerator.

Although our example may not exactly represent what is commonly sold, it is presented so that you can see just how much information can be obtained about how most appliances work if a schematic diagram is provided. In many instances, a schematic diagram can be obtained from the manufacturer if you simply request one.

At this point, you should have an initial working knowledge of electrical schematics and not be afraid to try to understand one you may come across. Many will be far more complex and contain symbols not mentioned in Chapter 6, but often they are self-explanatory, and symbols can always be looked up on the internet when necessary. Since the goal of this book is to give you a working knowledge of basic electricity, you may not be able to understand, at this point, how more complex electronic devices work, or those that use components with which you are not yet familiar.

Review of Chapter 7

- By carefully examining and tracing the connections on a schematic diagram you can usually figure out exactly what is happening in the electrical device you are interested in.

- Complex schematic diagrams can often be broken down into smaller sections that are easier to understand.

- If you ask them manufacturers of many electrical devices will send you a copy of the schematic diagram of an appliance that you own or they will tell you where to obtain one (usually by downloading one from their website).

VIII
Electrical Protection Devices

To complete our discussion of basic topics regarding electricity, it is now time to look at some common devices used to protect electrical circuits from overloads.

When electrical circuits are subject to a current flow that exceeds the capability of the interconnecting wires—such as a short circuit caused by crossed or damaged wires, or too many devices drawing power from the same supply—the system is commonly protected by a device called a fuse. In the past, a fuse consisted of a short length of wire designed to melt and interrupt the circuit when a predetermined amount of current flowed through it. Figures 28 and 29 show cross sections of an old-fashioned type of fuse as well as a couple of more modern types. As you may remember from Chapter 2, every conductor of electricity has an inherent resistance. When current flows through a conductor, some electrical energy is converted into heat. For most electrical wiring, the resistance is very low, so the heat produced is also very small, and the wire may only become slightly warm. In a fuse, however, a portion of the conductor is made from a material with a relatively low melting point. As a result, when the current exceeds a predetermined level, the heat produced is sufficient to melt the fuse element and interrupt the circuit.

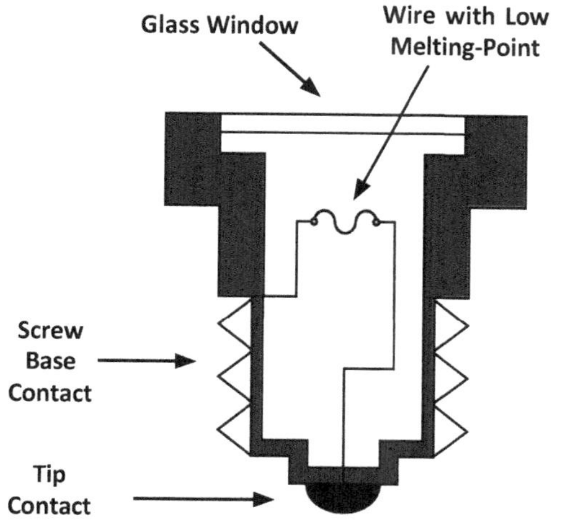

Figure 28, Cross Section of a Traditional Fuse

Fuses are almost always connected in the live wire of a circuit. They come in many sizes and shapes in addition to those shown here. All, however, use a conductor with a low melting point and operate in essentially the same way. Most fuses are labelled with the current rating at which they will operate and are single-use devices in that, once they have operated, they must be replaced.

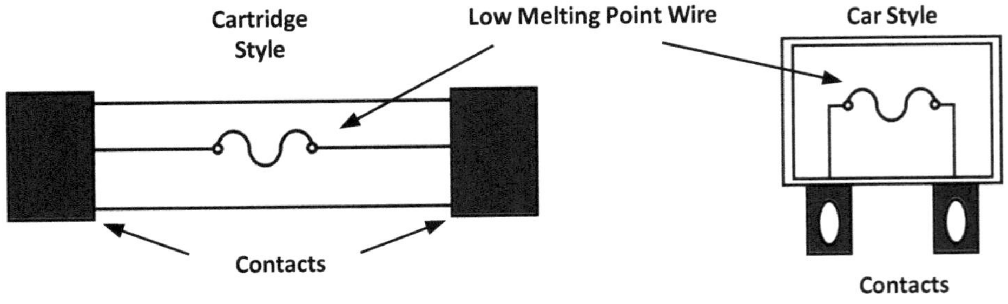

Figure 29, Cross Sections of Modern Fuses

In many cases, a fuse may provide adequate protection, but it is often desirable to protect a circuit without needing to replace the protective device after it operates. This is done using a device called a circuit breaker. One common type used in homes is the Miniature Circuit Breaker (MCB). Like a fuse, a circuit breaker will turn off a circuit when too much current flows, but unlike a fuse, it can be reset.

There are two main types of circuit breakers: those that must be reset when an overload occurs, and those that reset themselves when the overload is removed. The device shown in Figure 30 is typical of the ones found in the average home. As you can see, this one will **trip** when the current through it exceeds 16 amperes. At that point, the lever will move to the OFF position and will then have to be manually moved to reset it.

Figure 30, Typical Circuit Breaker

The internal construction is beyond the scope of this book, but there is usually an internal electromagnet that exerts enough force on a mechanism that pulls two internal contacts apart.

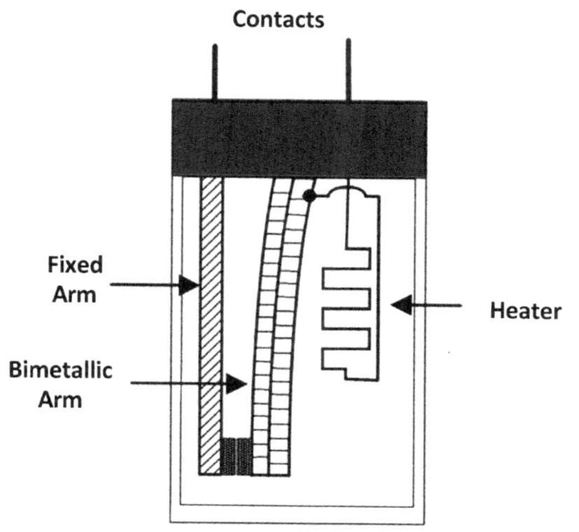

Figure 31, Typical Self-Resetting Circuit Breaker

The second type will not trip when the rated current is flowing, but will trip when the current reaches a predetermined overload point. When the fault is removed, the circuit will reset. Figure 31 shows how one style of this type works. This unit contains one fixed arm of rigid metal and a second arm with two different metals bonded together so that, when heated, the arm will bend. Adjacent to the bimetallic arm is a heater that is connected to the output terminal. With normal current flow, the heater does not get hot enough to bend the bi-metallic arm and the circuit remains closed. But when too much current flows due to an overload, the heater causes the bimetallic arm to bend away from the fixed arm and break the circuit. When this occurs, the heater is also cut off. As the heater cools, the circuit is closed again. If the fault continues, the cycle repeats over and over until the fault is corrected.

A third safety device is the Residual Current Device (RCD), often used as a plug-in adaptor or incorporated within a consumer unit (see Figure 32). This device is required by UK wiring regulations for many circuits, particularly in areas where there is a risk of moisture, such as bathrooms and kitchens. The device works by monitoring the current flowing in the live and neutral conductors. Under normal conditions, these currents are equal.

If they are not equal, it usually means that there is a leakage path (often to earth) somewhere in the device connected to the circuit, and the circuit is immediately disconnected. If this path happens to be through a person, there is a potentially dangerous situation that could result in a serious electrical shock.

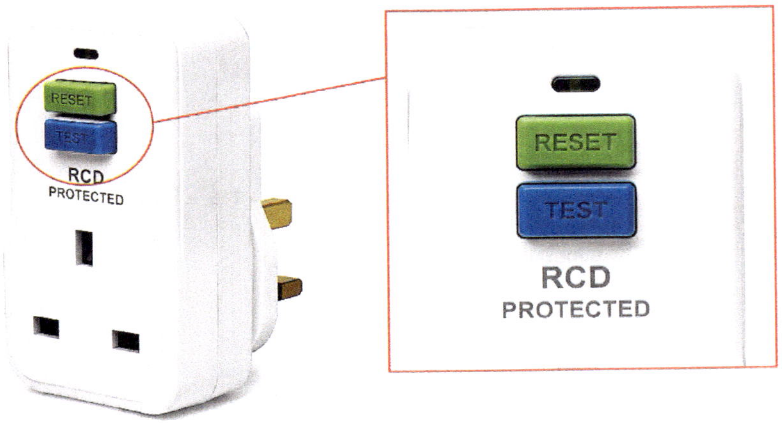

Figure 32, Typical Plug-In RCD Adaptor

The test and reset buttons on the RCD are used to check and restore its operation. Whenever you use or test any of these electrical protection devices, never attempt to bypass or defeat them. They are designed to protect the user and reduce the risk of fire.

Review of Chapter 8

• A standard fuse is used to protect an electrical circuit when too much current flows through it. It employs a special internal wire that melts when the current is too high, thereby opening the circuit and stopping the flow. This type of fuse cannot be reset and must be replaced.

• A circuit breaker (often referred to as an MCB in UK installations) performs the same function using an internal mechanism that operates when the current becomes too high. The breaker can be reset by simply switching it back on after the fault has been corrected.

• A self-resetting protection device also opens the circuit when the current is too high, but then resets itself when the current drops below the tripping point.

• *Never try to defeat or bypass any protective device. If it has operated, there is usually a good reason, and the problem should be corrected.*

IX
Basic Electrical Troubleshooting

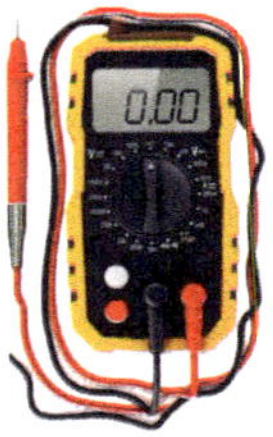

With your current understanding of some common electrical components (and taking the appropriate precautions) you should be in a good position now to determine what is wrong when an electrical appliance does not operate properly. To begin, you will need the multimeter we described in Chapter 4. We will then describe routine troubleshooting procedures you can perform using your multimeter. If the one you purchased is somewhat different, it should still be very easy to use because most low-cost units will be almost the same, although the labels may differ somewhat. Keep in mind that the purpose of this chapter is to give you an idea of what you can troubleshoot with your current knowledge. *Anything more involved than the following tests, or which require that you open a device, should be performed by experienced personnel.*

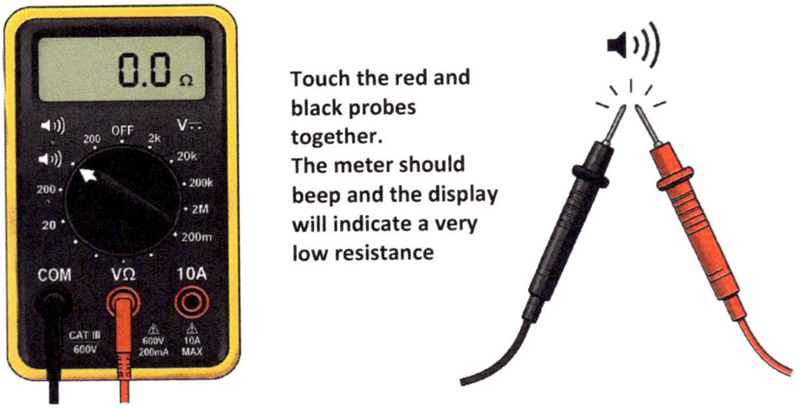

Figure 33, Multimeter Continuity Setting

Continuity Tests

Turn on your multimeter and set it to the continuity setting or to the lowest resistance (ohms) range, which is usually indicated by the Greek letter omega (Ω) and sometimes also with a sound or buzzer symbol. Touch the two probes together, and you should hear a beep, or, if your multimeter does not have this feature, the display will indicate a very low resistance value or zero.

This indicates that there is continuity between the ends of the two probes. You now have a continuity tester that can be used for many purposes.

If you wish to test an electrical switch, first ensure it is completely disconnected from any circuit. Then touch the two probes to the terminals of the switch and operate it. If the switch is functioning correctly, you will hear a beep (or see a very low resistance reading) when it is in the ON position. The same method can be used to determine the connections of multi-position switches.

Before attempting any of the procedures below, be absolutely certain that the device you are testing is not connected to any source of power.

To test an extension lead for breaks, connect the multimeter probes between each pin of the plug and the corresponding contact at the socket end. A beep or a very low resistance reading indicates a good connection.

To test an incandescent lamp for continuity, place one probe on the base contact of the (removed) bulb and the other on the metal shell. If the bulb is good, the meter will indicate continuity or a low resistance value. This test will not work properly with LED (light-emitting diode) or CFL (compact fluorescent) lamps, as they do not contain a simple filament.

Keep in mind that the above tests can be used to check for continuity in most situations where a break in a circuit is suspected.

Finally, if you wish to check an appliance, first ensure it is not connected to any source of power. You can then connect your multimeter, set to a resistance (ohms) range, between the pins of the mains plug. Turn on the appliance's power switch (while it remains unplugged). Depending on the design of the appliance, the meter may show a low resistance, a higher resistance, or even no reading at all. These results alone are not always sufficient to determine whether the appliance is functioning correctly. If a fault is suspected, further testing should be carried out by someone with more experience.

DC Voltage Test

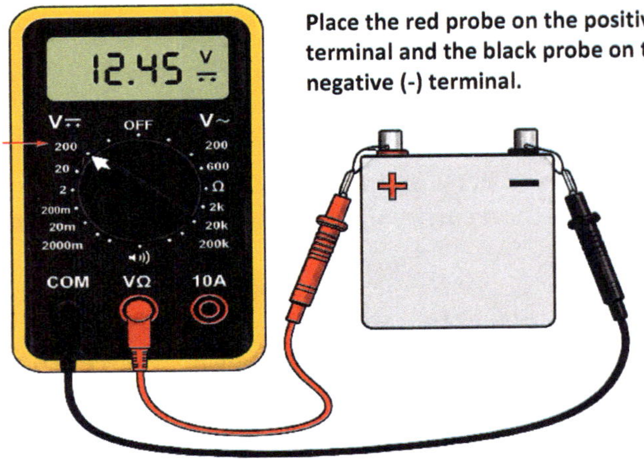

Figure 34, Measuring DC Voltage

To test a battery, first set the multimeter to its highest DC voltage range. Next place the red probe on the positive (+) terminal of the battery and the black probe on the negative (–) terminal. Slowly reduce the range setting until you obtain a reading on the multimeter. This reading represents the battery voltage (with no load connected) and will also confirm that the polarity of the terminals is correct.

The highest range is selected first in case the battery voltage is higher than expected. Remember that batteries range from as low as about 1.2 volts to as much as 48 volts or more in some cases. Using too low a range may result in an overload indication on the meter and, in some cases, may damage it.

To test a battery-powered appliance that is not working properly or operates intermittently, first examine the contacts in the battery holder for corrosion or poor contact pressure (usually at the negative terminal or spring contact). Then, with the batteries installed, turn on the appliance and connect your multimeter (set to a low voltage range) between the battery terminal and its corresponding contact, if accessible.

If the connection is good, there should be little or no voltage reading. Repeat this test for all battery contacts. A good connection will have minimal voltage drop, while a poor or corroded connection will show a measurable voltage. If a voltage drop is observed, turn off the appliance, remove the battery, clean the contacts, and retest.

AC Voltage Tests

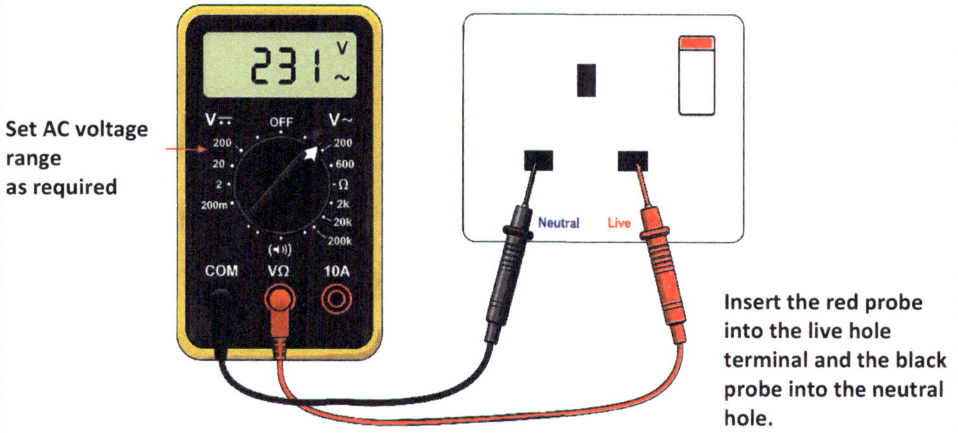

Figure 35, Measuring AC Voltage

To test for the presence of AC power at a mains socket, first set the multimeter to a suitable AC voltage range higher than the expected mains voltage. On many meters this will be the highest AC voltage range.

Carefully insert the probes into the appropriate socket contacts. Take care not to touch any exposed metal parts of the probes. The meter should indicate approximately 230 volts in the UK and many other countries. If the expected voltage is present, then power is available at the socket.

While performing this test, you can also check the earth connection. Place one probe in the neutral contact and the other in the earth contact. There should be little or no voltage present—typically only a few volts or less. Since neutral and earth are connected at the supply point, there should not be any significant voltage difference between them. To check an extension lead, it is safer to test it for continuity when it is not connected to any power source, as described earlier. Testing a live extension lead should only be carried out with appropriate care and experience.

Review of Chapter 9

- *A multimeter can be used to check continuity. This should always be done with the item being tested disconnected from any power source.*
- When making DC voltage measurements, select a range higher than the expected voltage, then switch to a lower range to obtain a more precise reading.
- When making AC voltage measurements, select a range higher than the expected voltage, then switch to a lower range as needed. Be sure to use the AC setting for AC measurements. Using the DC setting may result in an incorrect reading and, in some cases, may produce an overload indication on the meter.
- *When making AC voltage measurements be very careful not to touch the metal portion of the probe tips or the multimeter connections.*
- *When checking mains sockets, exercise extreme caution, as the voltage present is high enough to cause serious injury or death if direct contact is made.*

X

Summing It All Up

At this point, you should have a basic understanding of electricity and an appreciation of its dangers. While you should no longer be afraid of it, you should always have a great deal of respect for it. This chapter briefly summarizes the main points you have learned.

Fundamentals

- Like water, electricity also flows, but in this case it is the movement of electrons that produces the electrical current.

- A continuous flow of electricity requires a complete circuit from its source and back again. If this path is broken or interrupted at any point in the circuit, the electrons will not flow.

- Electrons will only flow through a conductor and are blocked by an insulator.

- Most metals and some liquids are conductors, while most plastic, rubber and glass items are insulators.

- An electric flow of electrons is kept inside a wire by an insulating coating around the outside of the wire.

- The amount of electrical current flowing in a circuit is measured in amperes while the pressure available to move or push these electrons is measured in volts.

- In order to have an operating electrical circuit you must have voltage and current (amperes). You can't have one without the other.

- Electrical power is a function of both voltage and current and is measured in watts. Watts are determined by multiplying the volts and amperes present in the circuit.

- High voltage with extremely low current will not necessarily be dangerous, just as high current without significant voltage will not necessarily be dangerous. However, never ignore any warning or caution indicator when using electricity, since you may not know all of the actual conditions present. You must fully understand the situation to determine the real level of danger.

- Current flowing in one direction only is called DC or direct current. Current that repeatedly changes direction is called AC or alternating current. Most electrical power to homes is AC.

Measuring Techniques

- A multimeter is an instrument used to measure voltage, current and resistance.

- When using a multimeter, first set the range higher than expected, then reduce it to obtain a suitable reading. Use the correct setting to avoid inaccurate readings or overload indications.

- *Never touch the metal portion of any probe tip while it is in use.*

- *Never work on a live circuit unless it is necessary for measurement. Mains electricity circuits have sufficient voltage to push enough current through the body to cause muscle contraction, ventricular fibrillation and other life-threatening conditions. DC circuits normally do not have high voltages present, but even relatively low voltages (a 12-volt car battery, for example) can produce high currents that may cause heating of wires and result in burns.*

- Always pay strict attention to high-voltage warning signs. Although voltage alone does not determine the level of danger, it can drive current through the body. Warning signs do not usually indicate all of the conditions present.

Figure 36, Common Electrical Warning Signs

The images in Figure 36 show some common high voltage warning signs. Whenever you see one of these or variations of them, be extra cautious.

Electrical Components and Details

- An electrical switch is used to make and break an electrical circuit.

- DC power connectors are often less heavily insulated than AC connectors, as DC voltages are usually lower. However, high DC currents can still cause wires to overheat and become very hot. When working with DC circuits, appropriate care should always be taken.

- The cable used to distribute electricity throughout a home usually contains three conductors inside the same outer sheath.

- In the UK, the live wire is typically brown, the neutral wire is blue, and the protective earth wire is green/yellow striped.

- The protective earth conductor is connected to earth via an earth electrode, typically a copper rod driven into the ground outside the building. If this connection is not present (as may be the case in some older installations), extra caution should be exercised.

In conclusion, it is the author's sincere hope that you now have a reasonable working knowledge of electricity and will not be afraid to discuss electrical projects with an electrician or even attempt, with caution, to troubleshoot and perform various simple electrical repairs around the home.

XI
Some Basic Electrical Repairs

This chapter presents realistic step-by-step examples illustrating how to solve specific problems.

Extension Lead Not Working or Intermittent

1. First, make sure that the wall socket is working properly—but do not plug anything into it just yet.

2. Now carefully place the probes of your multimeter into the socket end of the unplugged extension lead. Try to ensure that they are making contact with the internal contacts of the socket.

3. Next, set the multimeter to the highest AC voltage range and then, taking care not to touch the probes, plug in the extension lead. If the multimeter reads the correct mains voltage, then the extension lead is functioning correctly and the problem lies elsewhere.

4. If there is no reading on the multimeter, then while keeping everything plugged in, slowly bend the extension lead cable back and forth every few centimetres along its length to see if there is a possible break or loose connection in one or more of the wires inside the cable. If there is, then you will see the AC reading going on and off as you flex the cable. In this case, the extension lead should be replaced. Do not attempt to repair damaged internal cable wiring.

5. However, if there is no intermittent indication, unplug the extension lead from the mains socket and, using the lowest resistance (ohms) range of your multimeter, check continuity between each pin of the plug and the corresponding contact at the socket end. If an open circuit is detected, carefully open the connectors of the unplugged lead, one at a time, and look for a broken or damaged connection. These connectors can often be opened with a standard screwdriver. Since the wires are usually secured by screw terminals, they may simply need tightening. If the wire ends are frayed or damaged, they will need to be carefully re-stripped and then reconnected.

To strip a wire, use a wire stripper and remove approximately 6 to 12 mm of insulation. If the wire is stranded, tightly twist the strands together and then connect them, making sure no loose strands can come into contact with any other terminal.

6. If the extension lead is a fully moulded product, then the connectors usually cannot be repaired, but replacement connectors can often be purchased for both ends of the lead. You can then cut off the defective connector and replace it with a new one, which may save the cost of replacing the entire extension lead. If you replace a connector, make sure the wires are reconnected in exactly the same way as before.

Toaster or Electrical Heater Not Working Properly

Electric heaters and toasters are simple devices that contain an internal heating wire made from a material with high electrical resistance. As a result, during operation, the wire is designed to become very hot. If you have a defective heater that does not heat up, it means that there is a break somewhere in the circuit. To troubleshoot this, you must first determine how to safely open the unit to expose all of the internal connections. When doing this, note exactly what you have done so that you can properly reassemble the unit when the repair is complete. Next proceed as follows:

1. First, with no power connected to the unit, visually inspect all connections. Look for obvious loose wires or breaks in the heating element wire.

2. Using your multimeter set to the ohms range, place one probe on one pin of the mains plug. If the device has an operating switch or timer, set it to the normal "On" position. Using the other probe, slowly move from one connection point to another.

3. As you move the probe, you will reach a point where continuity stops. Note this point, then repeat the procedure from the other pin of the mains plug. The fault is located between these two points. If it is a broken heater wire, repair should only be done using a suitable high-temperature splice, such as those shown in Figure 37. If a splice is used, it should make good electrical contact and securely hold the wire ends.

4. If the break is at the operating switch or timer, then you need to either obtain a replacement part from the manufacturer or replace the appliance. Sometimes electrical connections to switches or timers are made with so-called **push-on terminals** such as those shown at the top of Figure 37. These are usually available at hardware stores or DIY centres.

Figure 37, Typical Crimp Terminals and Splices

Dealing with a Sudden Power Failure in Your Home

When you suddenly lose electrical power in your home, first check to make sure it is limited to your home and is not a problem affecting your whole neighbourhood. If you are not sure, you can phone your local electricity supplier. If it is just in your home, proceed as follows:

1. Open the door of your consumer unit or fuse box. Newer homes will have consumer units, while older ones may still have fuses.

2. Look for a blown fuse with a break on the inside element or a circuit breaker in the consumer unit that is in the tripped position. Figure 38 shows what either of these might look like. If you do see this, then the fuse must be replaced or the circuit breaker must be reset.

3. Before replacing a blown fuse, *first* try to locate the area or room in your home where the fault occurred. Correct the problem at that point by removing or unplugging whatever device caused the fault in the first place. Then remove the blown fuse and replace it with an exact equivalent. *Never use a larger fuse, as it will allow more current to flow, overheat the wires going to the faulty circuit, and in some cases even cause a fire.*

4. If your home uses circuit breakers, first try to locate the area or room in your home where the fault occurred. Correct the problem by removing or unplugging whatever device caused the fault. Now reset the circuit breaker by first sliding the switch to the OFF position, then sliding it back to the ON position as shown in Figure 38.

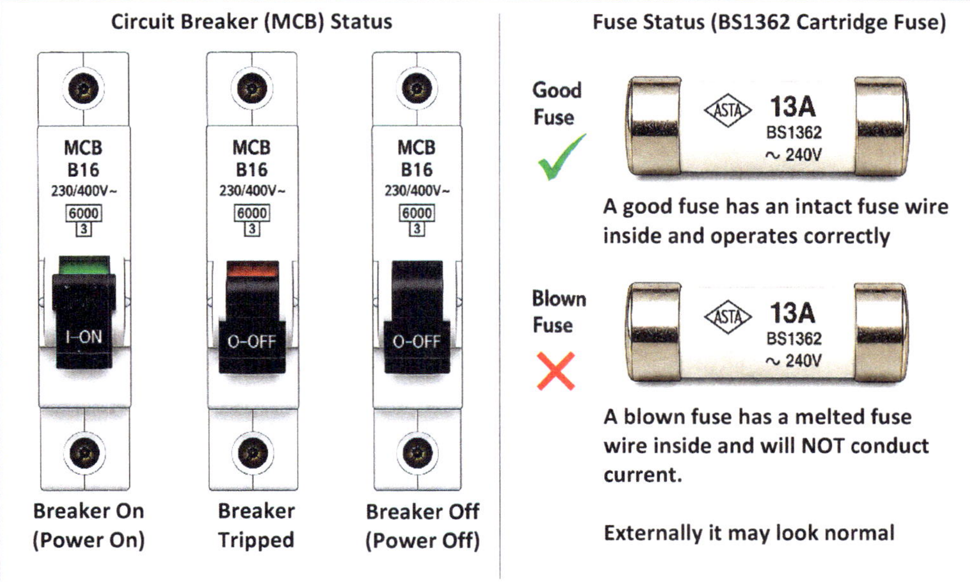

Figure 38, Fuse and Circuit Breaker Status

Replacing Defective Light Switches or Wall Sockets

Occasionally a light switch or wall socket will stop operating or become intermittent. When this happens, it should be replaced as follows:

1. *First, make absolutely certain that all electrical power going to the area or room where the defective switch or wall socket is located is turned off at the consumer unit or fuse box. Do this by switching off the appropriate circuit breaker or removing the fuse. This is for your safety!*

2. Remove the screws holding the defective switch or wall socket and carefully slide it out of the back box that holds it.

3. *Make a written note of exactly which terminal each of the wires going to the switch or wall socket is connected to so you do not forget! This is important since, when a switch is turned off, the live wire must not be connected directly to the appliance being controlled (such as a lamp) for safety reasons. This is also especially true of so-called two-way switches. If you connect this type of switch incorrectly, it may not work properly or at all. With regard to a wall socket, the live, neutral and earth wires must also be connected properly.*

4. *In addition, for safety, when replacing switches or wall sockets, be sure to always reconnect any earth conductor exactly as it was originally.*

5. Finally, slowly and carefully return the switch or wall socket to the box that originally held it. Make sure all wires are securely connected and are not pinched, which could cause problems in the future.

6. Turn the power back on and check that everything operates as it should.

The above examples show how you might go about troubleshooting a defective appliance or circuit. *In all cases, exercise extreme caution, especially when working with mains voltages.* Do not be afraid to attempt simple repairs, but carefully and within your level of experience. If you understood the examples in the previous chapters a great many such repairs will become straightforward. Items such as heaters, electric motors, and thermostats can be checked with your multimeter and, when you have done a few, you will become more confident in your abilities. In fact, one of the more frustrating parts of many repairs such as these is figuring out just how to open an appliance to get to the point where you can start making measurements.

Dealing with Emergencies

Throughout this book we have stressed the danger and cautions that should be understood and taken to protect you from the unfortunate accidents that a careless person could suffer when dealing with electricity. If such an occasion does occur the following is a short discussion of what must be done.

A Person Has Come into Contact with a Live Wire

- Do not touch the person with your bare hands!
- Immediately disconnect the power. If this cannot be done safely, push or pull the person away from the live wire using a non-conductive object such as a wooden or plastic handle, a wooden chair, etc.
- Call for emergency help.
- Under no circumstances attempt medical treatment beyond your level of training.

A Shock or Tingle Occurs When Touching an Appliance

Immediately unplug the appliance from the mains supply or, if this is not possible, switch off the appropriate circuit breaker or remove the fuse supplying the wall socket in the area to which the appliance is connected. When you are sure that all electricity is turned off, then you can proceed with the repair or replace the appliance.

An Electrical Device in Your Home Gets Excessively Hot

As in the previous case, immediately unplug the appliance or switch off the appropriate circuit breaker or remove the fuse supplying the wall socket in the area to which the appliance is connected. When you are sure that all electricity is turned off, then you can proceed with the repair or replace the appliance.

An Appliance Falls into a Sink, Bathtub, or a Container Full of Water or is Doused with Water

If the appliance is on, do not touch it. First, immediately disconnect the power at the consumer unit or unplug it from the mains socket if it is safe to do so. Allow the item to dry thoroughly for at least 24 hours. After that, a qualified electrician should carefully test it. In this case, you have no way of determining what internal damage to the appliance may have occurred, so be extra careful.

Sparks are Coming from a Mains Socket or Switch

Immediately switch off the appropriate circuit breaker or remove the fuse supplying the area in which the defective wall socket or switch is located. When you are sure that all electricity is turned off, then you can proceed with the repair or replace the item.

An Electrical Extension or Mains Lead is Getting Very Hot When in Use

If it is an extension lead, unplug it from the mains socket to which it is connected. Make sure that the extension lead is not overloaded by checking that its current rating is suitable for the connected appliance. If it is a mains lead attached to an appliance that is getting too hot, then examine the appliance for damage and, if necessary, repair or replace it.

Appendix

Note that all data is from general sources. Due to the wide variation among manufacturers from all over the world, the values here should be considered approximate. If details are required for a specific item then the manufacturer of that item should be contacted directly for exact values, or visit their website.

Suggested Copper Conductor Sizes for a 2% Voltage Drop at 230 V AC

Use these figures as a practical guide when choosing extension leads or cable sizes. All conductor sizes in this chart are given in square millimetres (mm^2), and all distances are given in metres.

Suggested Copper Conductor Sizes for a 2% Voltage Drop at 230 V AC

Maximum Current	Maximum Distance @ 230 V AC (conductor size – metres)
1 A	0.75 mm^2–100, 1.0 mm^2–130, 1.5 mm^2–200
5 A	0.75 mm^2–20, 1.0 mm^2–26, 1.5 mm^2–40
10 A	1.0 mm^2–13, 1.5 mm^2–20, 2.5 mm^2–33
13 A	1.5 mm^2–15, 2.5 mm^2–25, 4.0 mm^2–40
16 A	1.5 mm^2–12, 2.5 mm^2–20, 4.0 mm^2–33
20 A	2.5 mm^2–16, 4.0 mm^2–26, 6.0 mm^2–40
32 A	4.0 mm^2–16, 6.0 mm^2–25, 10.0 mm^2–40

Approximate maximum one-way cable lengths for copper conductors at 230 V AC and a 2% voltage drop.

(These figures are approximate maximum one-way cable lengths for copper conductors at 230 V AC and a 2% voltage drop. Always use cable that is properly rated for the current being carried and suitable for the intended application.)

In some older products aluminium wire may sometimes have been used. If replacing aluminium wire for any purpose always replace with copper!

Standard Domestic and International Wire Colour Codes

(These are the current colours used for most common AC electrical power wiring. Note that older (obsolete) colour coding may exist in some instances).

Wire Use	UK & Ireland	North America	Australia / NZ
Live	Brown	Black	Brown
Neutral	Blue	White	Blue
Earth	Green/Yellow	Green or Bare	Green/Yellow

The live wire carries current from the supply, the neutral wire returns current to the supply, and the earth wire is a safety conductor connected to earth. Additional wire colours may be used inside appliances for internal connections..

AC Power (Mains) Voltages

There are various standard AC voltages used for mains supplies in different parts of the world. The most common are approximately 120 volts AC at 60 Hz in North America and 230 volts AC at 50 Hz in the UK and most other parts of the world. Small variations in voltage are normal.

While you may not have any control over which standard is used in your area, keep in mind that equipment designed for one standard may not operate properly on another and could even be damaged if connected incorrectly. In addition, devices that depend on mains frequency (50 or 60 Hz) for timing, such as clocks and timers, may only operate correctly at the proper frequency.

Typical Wiring of AC Power (Mains) Plugs

There are many types of AC mains plugs used throughout the world, and standards may vary from country to country. The information shown in Figure 39 is based on common modern wiring practices but may not apply in all regions. In many cases the connections are polarised, so there is a specific indication of which wire must be connected to each pin.

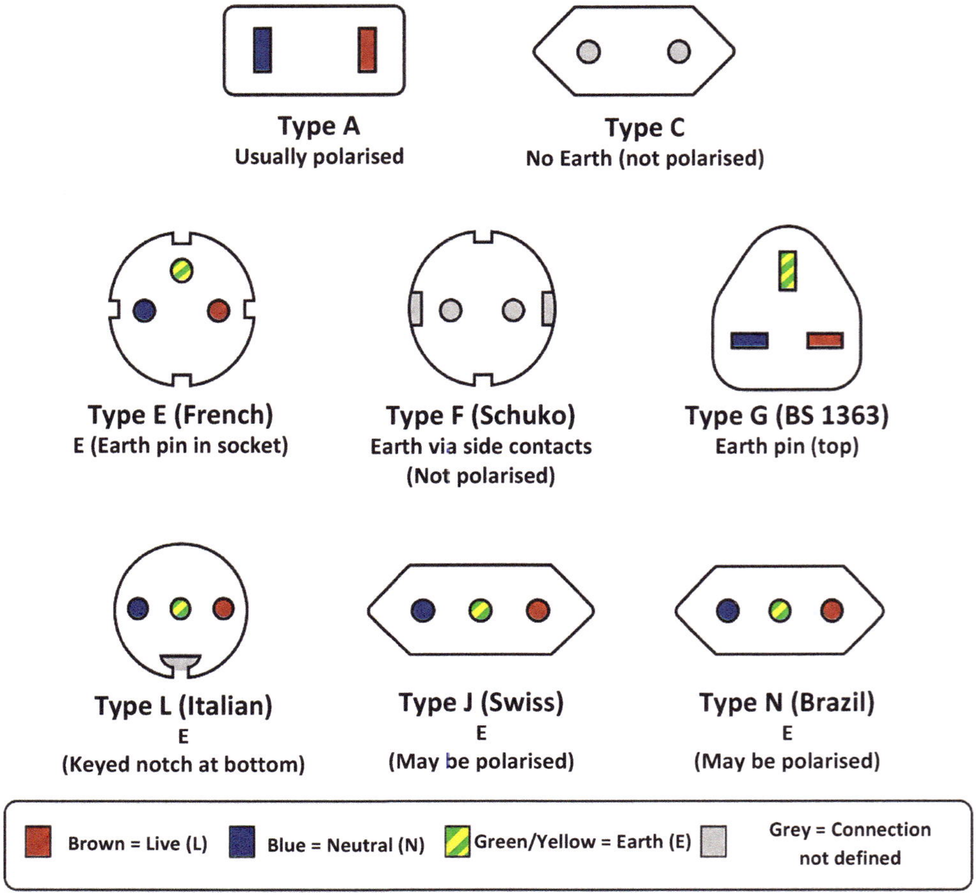

Figure 39, Common AC Mains Plug Connections (Typical Worldwide Configurations)

As a result, when using this diagram to replace a defective plug or wall socket, carefully note which terminal each wire connects to by its colour. Then reconnect the replacement plug or wall socket using the same wire colours and terminal positions. If the wires are the same colour, be especially careful to reconnect them to the same terminals as before. This will help prevent incorrect live and neutral connections..

In Figure 39, green/yellow represents the protective earth conductor, brown the live conductor, and blue the neutral conductor. Some plug types may not be polarised or may use different wiring conventions depending on the country and age of the installation. Always verify the correct connections before carrying out any repair.

Typical Common Dry-Cell Battery Ratings

The following chart shows some common household battery types, their typical voltages, chemistries and common applications. Actual capacities and performance vary depending on the manufacturer and intended use.

Typical Common Battery Ratings

Battery Type	Typical Voltage	Chemistry	Typical Use
AAA	1.5 V DC	Alkaline	Small portable devices
AA	1.5 V DC	Alkaline	Household devices
C	1.5 V DC	Alkaline	Medium-drain devices
D	1.5 V DC	Alkaline	High-drain devices
9-volt	9 V DC	Alkaline	Smoke alarms, testers

Common Rechargeable Batteries

Battery Type	Typical Voltage	Chemistry
AAA	1.2 V DC	NiMH
AA	1.2 V DC	NiMH
C	1.2 V DC	NiMH
D	1.2 V DC	NiMH
9-volt	8.4–9.6 V DC	NiMH

Battery capacities vary considerably between manufacturers and applications. Always use the correct battery type and polarity recommended by the equipment manufacturer.

Rechargeable batteries should always be charged using a charger designed specifically for the battery type being used. Follow the manufacturer's charging instructions carefully.

Typical Ratings for Small Electric Motors

(The following are typical motor sizes often found in small appliances such as fans, coffee grinders and pumps.)

Motor Horsepower	Approximate Watts	Typical RPM (No Load)	Typical Current at 230 V AC	Typical Current at 120 V AC
1/20	37	1550	1.25 amperes	2.5 amperes
1/12	62	1550	1.25 amperes	2.5 amperes
1/15	50	1550	1.40 amperes	2.8 amperes
1/10	75	1550	1.80 amperes	3.5 amperes
1/8	93	1725	2.00 amperes	4.0 amperes
1/4	186	1725	2.70 amperes	5.35 amperes
1/2	373	3450	4.90 amperes	9.8 amperes
3/4	560	3450	6.90 amperes	13.8 amperes
1	746	3450	8.00 amperes	16.0 amperes

Due to the wide variation in motors used in electrical appliances, the above ratings are only approximate and are intended as a general guide. If a motor requires replacement, always obtain the correct replacement type specified by the manufacturer.

Other Books by Irwin Math

Morse, Marconi and You: Understanding and Building Telegraph, Telephone and Radio Sets
ISBN 0-684-16081-1

Wires & Watts: Understanding and Using Electricity
ISBN 0-689-71298-7

More Wires & Watts: Understanding and Using Electricity
ISBN 0-684-18914-3

Bits and Pieces: Understanding and Building Computing Devices
ISBN 0-684-7879-6

Tomorrow's Technology: Experimenting with the Science of the Future
ISBN 0-684-9294-2

Author Biography

Irwin Math was the founder and Technical Director of *Math Associates Inc.*, one of the pioneering firms in fibre optic transmission technology and was formerly the president of *Liteway Inc.*, a manufacturer of fibre optic transmission systems.

He has been writing books, manuals, and technical articles since 1956.

For almost 45 years Mr. Math has contributed a monthly technical column called "*Math's Notes*" to CQ Magazine, an amateur-radio journal with a worldwide circulation of over 100,000 readers.

In past years he has produced various technical displays for leading museums in the United States and Canada including:
AT&T Infoquest, New York City, NY
Franklin Institute, Philadelphia, PA
Museum of Science and Industry, Ft. Meyers, FL
Ontario Science Center, Toronto, Canada
These displays are oriented toward explaining aspects of fibre optic technology to non-technical personnel.

Photo Credits

Lightbulb - Front Cover: ID 1358360405 © Cocomelo | Istockphoto.com
Laptop, figure 6 - page 16: ID 26119864 © Mpfphotography | Dreamstime.com
Hair dryer, figure 6 - page 16: ID 76770835 © papicasso | 123rf.com
Table fan, figure 6 - page 16: ID 61707674 © apopium | 123rf.com
Toaster, figure 6 – page 16: ID 88317023 © pattsuwat | 123rf.com
Electrical Terminals, figure 17 - page 31: ID 71179999 © Pichit Boonhuad | Dreamstime.com
Battery terminal connector, figure 17 - page 31: ID 64528727 © Peepooooo | Dreamstime.com
Generic digital multimeter, page 21 - figure 10: ID 163577721© destinacigdem | 123rf.com
Generic digital multimeter, page 25 - figure 12: ID 163575123 © destinacigdem | 123rf.com
Common Power Distribution Wire, illustration – top of page 30: with kind permission from **Nassau National Cable**
Plug for laptop, figure 17 - page 31: ID 189309831 © alexei88 | 123rf.com

Marvelous Magnetic Machines:
Building Model Electric Motors from Scrap

By **H.P. Friedrichs**

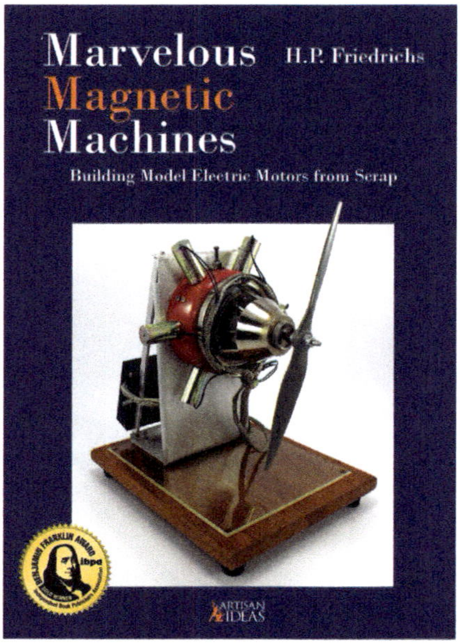

Gold Medal Winner, IBPA Award, Hobbies and Crafts

"I have a soft spot for things built from junk, and when those things happen to be precision electric motors I can't help but get extra inspired. In this book, H.P. Friedrichs breaks down the basics of how to build motors from things you can often find at garage sales and other second-hand markets. To really nail home how cool this is, simply look at the cover. Yup, that's a magnetic motor built around a kitchen pot." - **Make Magazine Review**

H.P. Friedrichs (author of *The Voice of the Crystal* and *Instruments of Amplification*) returns this time to explore the principles behind the operation and construction of five simple, yet impressive, model electric motors. You'll learn where to hunt for and find materials, and where to salvage suitable bearings. You'll know where useful solenoids can be extracted from scrap, and how to fabricate bobbins to wind your own. You'll learn how to time your motors, fashion a connecting rod, make a commutator from scratch, use a Hall effect sensor to detect magnet position, use a transistor as a switch, and much more.

Hardcover: ISBN 9781733325042, 160 pages, 177 photos and illustrations
eBook: ISBN 9781733325066

Printed and bound by CPI Group (UK) Ltd, Croydon, CR0 4YY
09/11/2026
14999540-0001